NATURAL HAZARDS

Other Meridian titles:

Robert H Fagan and Michael Webber, *GLOBAL RESTRUCTURING*

Jamie Kirkpatrick, *A CONTINENT TRANSFORMED*

NATURAL HAZARDS

DAVID CHAPMAN

**AUSTRALIAN
GEOGRAPHICAL
PERSPECTIVES**

Series editors
Deidre Dragovich
Alaric Maude

Melbourne

OXFORD UNIVERSITY PRESS

Oxford Auckland New York

OXFORD UNIVERSITY PRESS

Oxford New York
Athens Auckland Bangkok Bombay
Calcutta Cape Town Dar es Salaam Delhi
Florence Hong Kong Istanbul Karachi
Kuala Lumpur Madras Madrid Melbourne
Mexico City Nairobi Paris Singapore
Taipei Tokyo Toronto
and associated companies in
Berlin Ibadan

OXFORD is a trade mark of Oxford University Press

National Library of Australia
Cataloguing-in-Publication data:

Chapman, David M.
    Natural Hazards.

    Bibliography.
    Includes index.
    ISBN 0 19 553564 2.

    1. Natural disasters. 2. Emergency management.
    I. Title. (Series: Meridian, Australian
    geographical perspectives).

363.34

Typeset by Mackenzies, Victoria, Australia
Printed by SRM Production Services Sdn. Bhd.,
Malaysia.
Published by Oxford University Press,
253 Normanby Road, South Melbourne, Australia.

This book is dedicated to the thousands of brave men and women from all over Australia and the South Pacific who, in the catastrophic 1994 bushfires in New South Wales, gave selflessly of themselves, fought, and won.

Foreword

Australian geographers have produced some excellent books in recent years, several of them in association with the 1988 bicentennial of European settlement in the continent, and all of them building on the maturing of geographical research in this country. However, there is a continuing need for relatively short, low-cost books written for university students to fill the gap between chapter-length surveys and full-length books, to explore the geographical issues and problems of Australia and its region, or to present an Australian perspective on global geographical processes.

Meridian: Australian Geographical Perspectives is a series initiated by the Institute of Australian Geographers to fill this need. The term meridian refers to a line of longitude linking points in a half-circle between the poles. In this series it symbolises the interconnections between places in the global environment and global economy, which is one of the Key themes of contemporary geography. The books in the series cover a variety of physical, environmental, economic and social geography topics, and are written for use in first and second year courses where the existing texts and reference books lack a significant Australian perspective. To cope with the very varied content of geography courses taught in Australian universities the books are designed not as comprehensive texts, but as modules on specific themes which can be used in a variety of courses. They are intended to be used either in a one-semester course, or in a one-semester component of a full year course.

Titles in the series will cover a range of topics representing contemporary Australian geographical teaching and research, such as economic restructuring, vegetation change, natural hazards, the changing nature of cities, land degradation, gender and geography, and urban environmental problems. Although the emphasis in the series is on Australia, we also intend to produce some titles on Southeast Asia, using the considerable expertise that some Australian geographers have developed on this region.

We hope geography students will find the series informative, lively and relevant to their interests. Individual titles will also be of interest to students in related disciplines, such as environmental sciences, planning, economics, women's studies and Asian Studies.

While the primary aim of the series is to produce books for students, the topics selected deal with issues of relevance to all Australians. We therefore hope that the general reader will find some of the titles of interest, and discover that geographers have something distinctive to say on contemporary environmental, economic and social issues. As the books assume little or no previous training in geography, and attempt to avoid a textbook style, they should be readily understood by the general reader.

Deirdre Dragovich
University of Sydney
Alaric Maude
Flinders University

Contents

Figures

Tables

Preface

The air is at blood heat. Visibility is down to a few hundred metres, and from somewhere in the haze, there is the cacophony of sirens. Green plants shrivelled and died yesterday. Today, catastrophe comparable to nuclear war broke out. Australia's oldest and largest city is besieged by fire. Surface transport to the rest of the country is severed. Hundreds of homes have been destroyed, some within a dozen kilometres of the city centre. Emergency shelters and transport termini are packed with people dazed by shock and the loss of a lifetime's assets.

Apocalyptic? Yes. Fantasy? No, reality, as the worst Australian bushfire season in living memory reaches its climax. A tragic, yet fitting context in which to write the preface to this work on environmental hazards.

I have long been fascinated by the violent forces of Nature — all my life, I guess — and now wish to share some of my understanding of these forces, and of how we may live with them. In the pages which follow, you will find out about all of the important natural hazards — those which originate in our biological environment, in the air around us, the solid (but not always stable) earth beneath us, and in the waters of the globe. It would have been easy to write a doomsday volume filled with accounts of the many environmental disasters which have impacted the people of the world over the last decade, but I have chosen rather to try to give you an insight into the causes of the hazards — to answer the 'how?' and 'why?' questions. And in addition, I have shown how we may think systematically about management of natural hazards. The book may be read by itself, without reference to other sources. However, for the benefit of those readers who wish to make a serious study of part or all of the subject matter, I have endeavoured to provide as complete and up-to-date referencing as possible. Unlike some geographical topics, the field of natural hazards and their management is very broad, and in the preparation of this book, I have followed a fascinating intellectual trail through areas as diverse as astrophysics and zoology, or engineering and entomology. The signposts are there for you to do likewise, if you wish.

It is not possible in this short work to deal with the full range of environmental hazards. I have given you a basis for understanding the most important of them, and their management. Less important ones are omitted: I have not attempted to deal with hazards which have a vanishingly small chance of happening; I have not dealt with those which are largely the outcome of personal choice; I have not included natural hazards of very slow onset and low catastrophe potential; and finally, those hazards involving natural phenomena, but for which the exposure is quasi-technological, or hazards which are wholly technological in nature, demand a completely different treatment than is possible in this short work.

An example of a *very rare event* not discussed is death from meteorite impact. Meteors ('shooting stars') are relatively common, and one or two may often be seen by an observer outdoors on a clear night. Meteorites (those which actually fall to earth) are far less common, with only a few documented cases of meteorites having actually been observed to fall and subsequently retrieved. The probability of being hit by a meteorite during one's lifetime is of the order of 1 in 2 000 000 000 000!

It may be argued that we cannot 'blame' Nature for the consequences if we deliberately place ourselves at high risk, and any discussion of *natural* hazards must exclude those for which *considerable freedom of personal choice* is involved, such as those encountered in adventurous recreational pastimes, or in deliberate exposure to natural forces known ro be beyond one's skill level or endurance: for example, death of a mountain climber by exposure to extreme cold, death by drowning of a poor swimmer caught in a rip current at a beach, or skin cancer from excessive sunbathing.

Natural hazards of very slow onset and low catastrophe potential are those posed by phenomena such as swelling soils, for which ample warning time of the impending problem is available. Expansive soils present a hazard of slow, almost imperceptible onset. Virtually all soils will shrink or swell to a certain extent according to moisture content, but those termed 'expansive' undergo a significant change in volume (greater than 3%) due to the presence of active clay minerals such as bentonite or montmorillonite. Expansion and contraction of these soils can lead to the movement of the foundations of houses and other structures leading to costly cracking damage.

Technologically mediated natural phenomena present a huge variety of hazards. For example, extension of irrigation technology in some tropical countries has unintentionally led to a resurgence of malaria. The building of reservoirs, tanks and pits can create bodies of slow moving or stagnant water — ideal habitats for the breeding of the mosquito which transmits the malaria vector. And there are natural fibres, such as asbestos, which have been used by humans for more than ten thousand years, yet are hazardous. In ancient times people mixed them with clay before firing to strengthen and reinforce pottery vessels, and textiles that combined the fibres of flax and asbestos were known for their resistance to fire and decay. Asbestos fibres are used worldwide: as insulation, as reinforcement in concrete water pipes, and as the inert and durable mesh material used in some filtration processes, for example. Individual filaments of asbestos may be less than 0.01 mm in diameter. Broken filaments are carried in suspension as dust particles and lodge in the lungs, giving rise to a disease known as asbestosis, a debilitating and sometimes fatal lung disorder. Asbestos is today one of the most heavily regulated materials, but it is not the only inorganic fibre in common use — nemalite and tremolite have similar properties, and synthetic inorganic fibres abound. After more than a thousand years of use, asbestos is being replaced by other, often fibrous, materials. Glass fibres, for example, have become the insulation material of choice in construction. It remains to be seen whether the substitutes will be as successful, or more or less hazardous. We are certainly not going to do without fibrous inorganic materials nor expunge them from our environment.

Technological hazards range from events such as gas explosions or nuclear power plant failure, through chemical and road traffic hazards to electrocution of individuals in the home. The ten worst industrial accidents over the last half century have been responsible for the deaths of about 10 000 people, while the ten worst natural disasters in the same period have been responsible for the deaths of almost 3 000 000 people. People accept a higher threshold of severity with respect to natural disasters than for major industrial accidents. Five deaths in a flood would be perceived quite differently from five deaths resulting from a chemical explosion. The public is more forgiving of Nature than it is of culpable individuals, agencies or corporations. Generally, technological disasters claim fewer lives per event than natural disasters, but the social and economic repercussions are usually more severe than the death toll alone would suggest. Even industrial accidents that result in no identifiable deaths may have severe and extensive impacts. For example, in 1979 a freight train became derailed near Toronto, Canada. The train carried a mixture of commodities including styrenes, toluene, propane, caustic soda and chlorine. The subsequent explosions and fires from the derailed propane cars and the escape of chlorine led to the evacuation of a quarter of a million people in 24 hours and the closing of the area for almost a week. Property damage was slight (except for the train and the track), there were no deaths, and only a few minor injuries. However, the dislocation of a quarter of a million people from their homes, and the disruption to business and to public sector activity led to costs associated with the event being estimated at over $70 million.

Finally, as we enter the world of the third millennium, the distinction between natural and technological hazards is in some cases becoming blurred. For example, record floods at Times Beach in Missouri, USA, created a particularly complicated disaster situation: numerous surface sites in the area had previously been contaminated with the chemical dioxin and the effects of flooding included toxic chemical contamination. The disasters at Chernobyl and Bhopal demonstrate the complex interactions possible when artificial toxic substances are released into natural atmospheric movements. And the impact of earthquakes, hurricanes, tornadoes and lightning on facilities such as power stations can be disastrous, leading to enormous secondary complications.

■

Introduction — Natural hazards and us

WHAT IS A NATURAL HAZARD?

Bushfires in Australia, floods in China, hurricanes in Haiti, and earthquakes in Peru: these are natural phenomena, awesome and wondrous in themselves, but capable of inspiring fear and dread if human life or property is affected by them. Only in the latter context may they be regarded as hazards: a tornado pirouetting across the Australian outback may well go unnoticed, and many a tropical cyclone observed by weather satellite never influences human life or property. The recognition of a natural phenomenon as a hazard also has to do with the degree to which the event exceeds normal human expectations at that place and time: a blizzard in winter at Barrow, Alaska, would be uncomfortable but unremarkable, while a blizzard in Sydney, Australia, even in winter, would create havoc — traffic chaos and accidents, frozen and burst water pipes, deaths from hypothermia (especially among the homeless), and an immediate energy crisis as people attempted to keep warm in a city totally unused to such weather.

A natural hazard may be defined as an *interaction* between a system of human resource management and an extreme or rare natural phenomenon, which may be geophysical, atmospheric or biological in origin (table 1), greatly exceeding normal human expectations in terms of its magnitude or frequency, and causing a major human hardship with sig-

Table 1 A taxonomy of natural hazards

Most natural disasters can be classed as originating in the solid earth (the lithosphere), the fluid earth (the atmosphere and hydrosphere) or the biosphere, although some events cut right across such boundaries, and interactions create secondary effects.

Hazards originating primarily from the fluid earth (the atmosphere and hydrosphere)

- Tropical and extra-tropical cyclone
- Tornado and other windstorm
- Thunderstorm and lightning
- Hydrometeors (rain, hail, freezing rain, blizzard, snow, sleet, whiteout)
- Flood (whether from rain, snow-melt or natural dam-burst, including secondary effects such as waterlogging and unwanted sedimentation)
- Storm surge
- Heat wave and cold spell
- Frost
- Fog
- Coastal erosion and other effects of extreme wave actions
- Rip-current
- Sea ice
- Sea level rise (low catastrophe potential)
- Drought

Hazards originating primarily from the lithosphere

- Earthquake (ground rupture, fault displacement, fault scarp movement, soil liquefaction)
- Volcano (explosion, lava flow, tephra fallout, ballistic projectiles, pyroclastic density current, lahar, toxic gas, atmospheric shock wave)
- Mass movement (including mudflow, soil creep, avalanche, debris flow, rockfall, landslide)
- Tsunami (primarily from earthquake or volcano, secondarily from the hydrosphere)
- Duststorm
- Swelling soil (low catastrophe potential)
- Sand drift (low catastrophe potential)

Hazards originating primarily from the biosphere

- Wildfire
- Bacterial, viral or protozoan hazards:
    directly affecting humans: bacterial, protozoan or viral diseases
    indirectly affecting humans, e.g., foot-and-mouth disease of cattle, tuberculosis of cattle
- Microflora:
    directly affecting humans, e.g., fungal diseases of the body, such as athlete's foot
    indirectly affecting humans, e.g., wheat-rust, potato blight, Dutch elm disease
- Macroflora:
    directly affecting humans, as from ingestion of poisonous plant materials, or allergic reactions to plant materials, esp. pollens

(continued)

Hazards originating primarily from the biosphere

> indirectly affecting humans, as from weed infestations, esp. outbreaks of alien plants

- Microfauna:
  directly affecting humans, e.g., mites or lice, or internal parasites of humans, such as schistosomiasis
  indirectly affecting humans, e.g., parasites of cattle
- Macrofauna:
  directly affecting humans, e.g., sharks or terrestrial carnivores; venomous snakes, fish, insects, or arachnids
  indirectly affecting humans, e.g., plagues of rodents, rabbits or locusts

nificant material damage to infrastructure and/or loss of life or disease. A particular level or severity of a natural event becomes a hazard only in relation to the capacity of society or individuals to cope. Gilbert White, pioneer researcher of natural hazards, observed that 'by definition, no natural hazard exists apart from human adjustment to it' (White, 1974: 3). However, although natural hazards are related to specific natural phenomena, there is no question that many are exacerbated, in magnitude or frequency, by human intervention: clearing of forest, or urbanisation, will change the behaviour of floods from a catchment, for example, and there is no doubt that many wildfires are lit by humans, either deliberately or unintentionally.

Most natural hazard events are accompanied by the violent release of energy, far exceeding our normal human experience or indeed the capacity of human beings to modify the environment (compare figure 1). As such, many such natural hazard events are associated with what has come to be known as *threshold* behaviour in natural systems: environmental change may be dominated by the effects of rare, catastrophic events, reflecting the truth of Francis Bacon's observation that 'in Nature, things move violently to their place, and calmly in their place' (cf. Coates and Vitek, 1980; Baker and Pickup, 1987; Harris, 1990; Johnston, 1992).

NATURAL HAZARDS IN HISTORY

There is no doubt that the course of history has been affected on numerous occasions by natural disasters. Take, for example, the failure of the Mongol invasions of Japan. Khublai Khan, the great Mongol ruler of the thirteenth century, attempted two invasions against Japan in 1274 and 1281 AD. In both cases his large armadas were heavily damaged, and the invasions failed as a result of severe storms at sea which historians agree were what we would today call typhoons or tropical cyclones (Neumann, 1975). In the summer of 1657 Denmark launched military action against

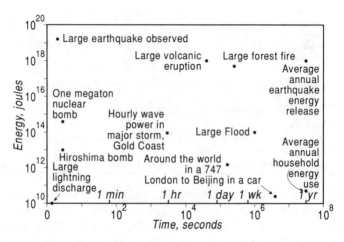

Figure 1 Comparison between the time–energy relationships of common natural hazard phenomena and powerful or energy-demanding human activities. The standard unit of measurement for energy-work is the joule: one watt, the normal measure of electricity consumption, is equal to one joule per second.

Sweden. Charles X, King of Sweden at the time, conquered Jutland, the westernmost part of Denmark, in November 1657. However, in the absence of an adequate naval force, Charles X could not carry his campaign to Zealand, the island on which Copenhagen is situated. Unexpectedly, the severe winter of 1657–58 came to his aid. By February 1658 the seas between Jutland and Zealand became frozen over, enabling the Swedish army to march over the ice from Jutland to Zealand and force the Danes to sue for peace, in what has been referred to as the 'most staggering exploit in all the history of Sweden' (Reddaway, 1952; Neumann, 1979). Exceptionally severe winter weather has also been implicated as a significant contributory factor to the 1789 Revolution, a watershed in the history of France (Neumann, 1977); and in more modern times, the creation of Bangladesh, as a national entity distinct from Pakistan, was an outcome of the impact of a tropical cyclone in 1970.

In the second half of the twentieth century, natural disasters have caused widespread damage and severe economic losses, and claimed the lives of over 4 million people around the world, while perhaps 1000 times as many have had their lives adversely affected in some way or other (see figure 2). The great devastation and loss of life which can be caused are highlighted by events such as the 1991 eruption of Pinatubo in the Philippines, with its dramatic and disastrous consequences on the economy, principally as a result of volcanic ash fallout, or the 1991 tropical cyclone in Bangladesh in which 138 000 people were killed. Berz (1991) published a table of losses from major natural disasters from 1960 to 1990 which totalled some US$140 billion, or about US$30 for every man, woman and child alive on the earth in 1991. Major natural hazards do and should receive considerable attention, but numerous small-scale events go unnoticed by the media, and many natural disasters occur in remote areas or closed societies where communications are poor and news travels slowly. For these reasons and others, knowledge of natural disasters, their severity and extent, their impact in terms of lives lost and property damage is neither easy to collect nor particularly precise when it has been assembled. The ten worst natural disasters in the last half century have accounted for about two-thirds of the 4 million death toll. The reported number of natural disasters worldwide that resulted in 25 or more deaths has fluctuated from year to year but generally followed an upward trend (figure 3). The trend *may* reflect an increase in the level of geological and meteorological activity, although it seems unlikely that the number of hurricanes, earthquakes, volcanic eruptions and so on would

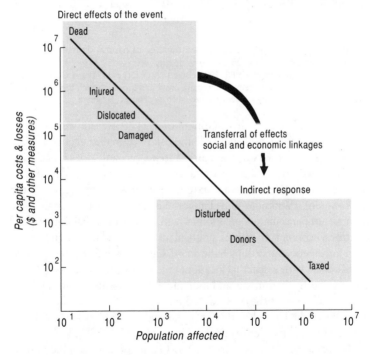

Figure 2 The impact of disaster exhibits a continuum of effects, and most people in a society will be affected in one way or another.
Source: redrawn from White and Haas (1975). Reproduced by permission.

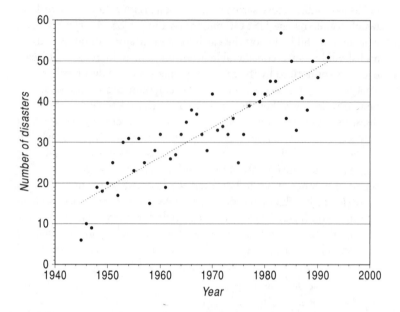

Figure 3 The worldwide trend in the number of natural disasters reported which were responsible for 25 or more deaths.
Data principally derived from UNEP (1991), Berz (1991), Degg (1992) and Glickman et al. (1992). Database excludes events of uncertain onset and those in which the deaths from hazard were not marked by sudden fatalities, such as drought or famine.

increase worldwide in this manner. In part, the trend results from improved reporting, especially the inclusion of smaller events in more remote regions. However, the upward trend certainly does reflect substantial growth in world population, the increasing vulnerability of marginal groups and, to some extent, the mismanagement of the environment (Wijkman and Timberlake, 1984). In the low-income countries in particular, population growth, increasing urbanisation, land shortages, economic hardships, political instability and migration to cities often force people to occupy hazardous locations, while endemic poverty limits choice with respect to location, emergency preparedness, and the ability of both government and individuals to respond to natural disasters. Recognition of these trends led to the designation by the General Assembly of the United Nations of the decade of the 1990s as the International Decade for Human Disaster Reduction (UNDRO, 1988).

Meteorological disasters are the most common catastrophic natural hazards and about 1000 such events have claimed the lives of over 1 million people since 1945, while the less frequent geological disasters have claimed rather more lives. Other natural hazards originate in the biosphere. Locust plagues and attacks from large carnivores are obvious biological hazards, but there are also less obvious phenomena. In Australia, disasters of meteorological origin are the most costly — droughts, storms (including tropical cyclones) and floods account for the greater part of the nation's $1.3 billion average annual damages bill.

THE PSYCHOLOGICAL IMPACT OF NATURAL DISASTER

Not all individuals view hazards alike; neither do all cultural groups. Beliefs about the 'purpose' of the natural environment with respect to human beings, and about the degree to which human activity can (or should) affect the nature of the environment differ widely, as do attitudes towards the environment as benign or otherwise, and the locus of responsibility for one's well-being. Tuan (1974) has commented that 'wherever we can point to human beings, there we point to somebody's *home*, with all the kindly meaning of that word,' and residents of environments fraught with hazard may nonetheless perceive their *home* to be safe and benevolent, not on the basis of objective evidence, but on the basis of normative beliefs about the nature of environment.

Recognition of the importance of the perception of hazard by people, and the perception of the range of possible responses, has led to greater attention being given to psychological concepts and techniques in hazard management. Concepts such as perception, attitude, motivation and personality have been examined, although few studies address the long-term psychological effects of disaster. Descriptive studies and clinical experience suggest that even the most horrifying and traumatic experience is organised in a person's mind (Raphael, 1979). Perhaps surprisingly, there is evidence to suggest that participants in disaster education programs experience higher levels of post-disaster stress than the general public (Faupel and Styles, 1993).

There are three main classes of threat or trauma to be considered: death/destruction; loss; and dislocation/relocation (Raphael, 1979). Death/destruction threats and traumata are experienced when the victim's life or body is threatened by the physical forces of the disaster and its consequences, or the individual is exposed to the death/destruction of others' bodies in massive or traumatic ways. The outcomes are often condensed in the traumatic 'screen memories' of the disaster, such as the vivid descriptions of overwhelming scenes of death and destruction recounted to Erikson (1979) by survivors from a catastrophic flash flood.

This area of threat/trauma is identifiable from the earliest stages after the disaster experience, unlike that of loss.

Loss threats and traumata occur following personal or property loss. The person may have been bereaved by the death of the primary attachment figure or a close family member, or someone more distant but still emotionally important. The grief may be on account of the many deaths of strangers, or a beloved pet. There may be loss of valued personal possessions with strong symbolic emotional attachment, or of property, home and means of livelihood. In the early phase after a disaster, losses are gradually identified as the 'inventory' process occurs. The euphoria of survival, or shock and denial, may mean that responses to the losses are not evident, however, and may not be for some weeks. The ensuing deprivations are increasingly experienced as time goes on.

Dislocation/relocation stress/trauma is associated with dislocation of the person from his family and/or natural environment including home and neighbourhood, and subsequent relocation. While separation and dislocation may occur acutely and transiently, they may also be very prolonged and contribute significantly to the chronic stresses of the post-disaster period. Spontaneous comments by many disaster victims indicate the severity and unpleasantness of this ongoing component.

There is also the device of investing the hazard with a personality, in order to make it easier to cope. Barker and Miller (1990) relate how anthropomorphisation of a hurricane formed part of the coping mechanism of the population of Jamaica in response to Hurricane Gilbert, the most intense tropical cyclone ever recorded in the southwest Atlantic. The practice of anthropomorphising nature is not new (Tuan, 1979), and it is possible that meteorological agencies encourage the process in the case of tropical cyclones by assigning personal names to them, but Barker and Miller argue that the anthropomorphisation of the hurricane provided the people of Jamaica with a way to reduce the awesome and incomprehensible to something prosaic and simplistic. It also provided the community with a new temporal and social reference point, and the terms 'BG' and 'AG' (i.e., before and after Gilbert) entered the patois. Moreover, personifying the natural event enabled it to become the butt of humour, an important psychological coping mechanism. A subsequent tropical cyclone named Joan achieved notoriety as Gilbert's wanton wife, searching for her husband, and the Jamaican *Daily Gleaner* reported that 'Gilbert, in his violence, wrote on the chalkboard of the Jamaican classroom the lesson that affluence and poverty have common cause when it comes to nature.'

COPING WITH NATURAL HAZARDS

While technological hazards are preventable, in theory at least, if people are willing to sacrifice the benefits offered by the technologies which create the hazard, we cannot presently prevent most geophysical or atmospheric hazards. Biological hazards also remain with us, but they are rather more subject to control, although the emergence of drug-resistant malaria has reduced our ability to control that disease in many developing countries. In these regions, the diarrhoeal diseases are also responsible for the deaths of millions of children annually, and traditional diseases such as tuberculosis are by no means conquered (Walsh and Warren, 1979).

With regard to recovery, Hirshleifer (1987) concluded from a study of great disasters that, in the instances examined, rather prompt recovery (say, within four or five years) to pre-disaster levels of well-being was technologically possible and did, in fact, take place with some 'slippage' due mostly to avoidable mistakes in monetary policies. For example, his analysis of the Black Death of 1347–51 in western Europe, which swept away over one-quarter of the entire population, shows that the record of that event provides no support for contentions that social collapse or an economic downward spiral are necessary or likely consequences of massive disasters. The short-term economic aftermath of the Black Death was very much in line with what economic theory would predict: a rapid rise in wages and per capita incomes of the labouring classes, and downward pressure on rents and the incomes of the propertied classes. The attempt to stem these pressures by government fiat had but limited success. Economic recovery was rapid. Although there are literary reports of organisational breakdowns in cities during the period when the plague was at its height, it is evident — from records of governmental proceedings, and the fact of large-scale military and naval activity — that the mechanisms of government did not collapse ('government' in the fourteenth century was, of course, a simpler and more limited activity than government today). The Black Death accelerated the decline of feudalism and the shift to modern contractual economic relationships. Although this development was proceeding in any case, the sudden relative scarcity of labour dictated a 'new deal' that the tradition-bound feudal system was unable to provide.

Biological adaptation to environmental factors such as high altitude is well documented, and it is possible that certain natural hazards may result in biological adaptations at the individual level, if the hazard is of sufficiently long duration. Cultural adaptation to hazards also occurs, with the cultural modifications mediating the effects of natural disasters and disease. Adjustments are motivated by the desire either to prevent the natural event or eliminate or mitigate its consequences. The degree to which a society is adapted to a particular hazard depends, among other things, not only on the information available about the likely magnitudes of hazard events and their potential impact, but on the willingness and ability of people to act on that information. Future losses may be prevented or ameliorated by present investment in hazard management, but the funds and/or motivation may be lacking, either at the individual or community level. Some hazards are successfully addressed at the individual level (for example, there is much that a householder can do to reduce bushfire hazard in the immediate environment of the home), but others only at a community or regional level (e.g., flood mitigation), and these may be neglected by a government obsessed with military defence, say, at the expense of civil defence.

Unlike countries where para-military organisations or defence forces are used as the front line of response in disaster emergencies, the Australian system for responding to natural disasters consists largely of volunteer based state counter-disaster organisations, such as the state emergency services, bushfire brigades, St John Ambulance, and rescue services, etc., along with police, regular fire brigades and ambulance services (NDO, 1990). There are about 400 000 members of volunteer emergency organisations, compared with approximately 65 000 paid personnel in police, ambulance and fire services. Although co-ordinated from state and regional offices, state emergency service units are usually closely linked with local government, and may utilise its resources. Military personnel are sometimes used in a support role. The Australian Constitution does not delegate specific powers concerning disaster relief to the federal government: these powers belong to the states. At the federal level, there is Emergency Management Australia (formerly the Natural Disasters Organisation), formed in 1974, almost on the eve of cyclone Tracy, with the objective of co-ordinating the resources of the federal government in the event of large-scale natural disaster (Stretton, 1979). Natural disaster-relief arrangements also exist whereby, in any given year, states are expected to meet a certain base level of expenditures for disaster relief from their own resources, and the federal government provides reimbursement of a proportion of 'eligible' expenditures above the threshold. Prior to 1989, drought stimulated more payments from the scheme than any other single hazard, but in 1989 federal government decided to exclude drought relief from its provisions.

■

Biohazards —
All creatures great
and small

Throughout most of the time humans have been on the face of the earth, it might be said that *Homo sapiens* was an endangered species. To the geophysical and atmospheric hazards discussed elsewhere in this book must be added the hazards posed by organisms both large and small. Organisms too small to be seen have been hazardous to humans throughout all of history. Man the hunter was himself also liable to be taken by large carnivores, and, when in search of small game or vegetable food, the disturbance of deadly snakes, insects, arachnids, or sea creatures posed the threat of severe discomfort, if not death. And when the hunter-gatherer found some attractive-looking vegetable food, was it safe? Many plants bear poisonous fruits, and toxic extracts from sap, bark, or other plant parts have been used for hunting, murder, and mayhem before history began. Who knows how many people succumbed to the toxins in plant tissues before some sort of systematised knowledge of their properties developed? Allergic reaction to plants such as poison ivy has long been recognised, but more recently, and especially in Australia it seems, we have become concerned about afflictions of our respiratory organs, such as hay fever or asthma, in which airborne plant materials are implicated. Today we face the threat of lethal, fast-spreading biological hazards resulting from biotechnology gone wrong (Bradford, 1993). Space does not allow even a catalogue of all the biological agents hazardous to humans: this chapter presents a selective overview.

BACTERIAL, VIRAL, OR PROTOZOAN HAZARDS

Most of our understanding of natural hazards comes from the study of geophysical events which constitute only one set of the phenomena which may be perceived as natural hazards. The biological hazards have remained virtually unexplored, and explicit mention of disease as a natural hazard is rarely seen, and if so only in passing (Lewis and Mayer, 1988). Bacteria, viruses, or protozoa of many kinds are always present in our immediate environment, often in huge numbers. When do they become hazards? Many may adversely affect humans, but those which satisfy the hazard criteria of rapid onset and high magnitude of impact are the *plagues* or *epidemics* such as bubonic or black plague, cholera, typhus, malaria, influenza, and some sexually transmitted diseases, especially AIDS. The ease of spread of influenza, from person to person, and worldwide by infected air travellers, makes it a hazard with the potential of a very rapid onset. The influenza pandemic of 1918–19 was responsible for upwards of 20 million deaths worldwide, and in terms of sheer numbers was the most serious epidemic which the world has experienced in the twentieth century. In the 1970 pandemic there were about 2 million cases in Australia alone.

Bubonic or black plague, often simply called the Plague, is an acute infection, primarily of rats and other rodents, secondarily of humans. The infective organism, *Bacillus pestis*, is spread by rodents and transmitted by their parasitic fleas to humans. The disease is marked by fever, chills, severe prostration and swelling of the lymph nodes in the groin and other parts of the body. The swelling in the groin may attain the size of a large orange and gives rise to the name 'bubonic plague' from the Greek *boubon* meaning groin. Furthermore the bacillus also causes haemorrhages which are called plague spots when they occur in the skin. The dark colour of these spots coupled with the extremely high mortality resulting from the disease gave it the name 'Black Death' in the Middle Ages. An outbreak of the Plague in 1347–51 must rank as one of the greatest disasters of all time, and gave the name of the era the 'Age of Calamity' (Time-Life, 1989). The disease spread right across Europe and Asia, completely depopulating 200 000 settlements in Europe, and killing between a quarter and a third of the population of the affected continents. At the turn of the twentieth century, the Plague was severe in the western Pacific, and reached Sydney, carried by rats on ships, in 1900. Quarantine of the victims, and a war on waterfront rats, curbed the spread of the disease in Australia: there were 1212 cases in the outbreak, which lasted until 1909, and 311 died.

Cholera is an acute diarrhoeal disease of the gastrointestinal tract caused by ingestion of the bacterium *Vibrio choleriae*. The incubation period ranges from a few hours to five days after the bacteria are ingested. If the disease is untreated the victim's chances of survival are less than 40 per cent, but in cases where patients receive proper treatment the chance of survival is greatly increased. Cholera epidemics are not part of a distant and medically primitive past: an epidemic in India in the 1970s claimed at least 5000 lives, and thousands also died in a Latin American outbreak of the disease in 1991.

Bacterial, viral, or protozoan hazards may also indirectly affect humans, in the form of diseases of livestock on which we depend for food or fibre (e.g., foot-and-mouth disease of cattle, tuberculosis of cattle) or animal diseases which may be directly transmitted to humans (e.g., rabies). Foot-and-mouth disease of cattle is an acute, highly contagious disease of cattle, sheep, and other cloven-footed animals. The virus affects the tissues of the mouth and feet, creating painful fluid-filled vesicles that severely damage the lips, tongue, gums, and feet. The affected animal cannot eat, and becomes lame. Control is costly and difficult, as the virus may be spread by human handlers of cattle (even on their clothing), by vehicles used for transport, and so on. Slaughter of affected animals, destruction of clothing known to be in contact with them, and subsequent quarantine of outbreak sites for some time afterwards are some of the measures which must be employed. Australia is free of the disease, but its potentially disastrous effects on Australian livestock industries is one reason for very strict quarantine laws. Emergency plans, and stockpiles of necessary supplies, including vaccines, also exist.

MICROFLORA

Microflora may directly affect humans, as by fungal diseases of the body, such as athlete's foot, but the most serious hazards created by them are indirect. Through their impact on important crops, or attack on other valued plants, microflora may indirectly affect humans in significant and sometimes life threatening ways. Plant diseases such as wheat-rust, potato blight, and Dutch elm disease are all caused by microflora. The Irish famine caused by potato blight in the mid-nineteenth century is probably the only major famine due to crop disease. Over a million Irish died, and several times that number emigrated (principally to North America) in the period, effectively halving the population of the island. Dutch elm disease is caused by a fungus (*Ceratocystis ulmi*) spread to trees of the genus *Ulmus* by bark-boring beetles. An aggressive strain of the disease introduced to Britain in the late 1960s destroyed over 30 million elm trees, completely changing the landscape of much of southern Britain (Jones, 1981).

Macroflora

Macroflora, or large plants, may directly affect humans from ingestion of poisonous plant materials (e.g., from the old world: belladonna or deadly nightshade *(Atropa belladonna)*, all parts of which are poisonous, especially roots and seeds; or ivy *(Hedera helix)*, which bears poisonous berries. Even the tubers of the humble potato *(Solanum tuberosum)*, which, incidentally, is related to belladonna, become poisonous when greened on exposure to sun, and an African member of the same genus, *Solanum aculeastrum*, is commonly known as poison apple owing to the toxic nature of its fruits at all stages of development. Introduced, and common in Australia, are oleander *(Nerium oleander*, from the Mediterranean region), all parts of which are poisonous if eaten, and the castor oil plant *(Ricinus communis*, from Asia), which bears poisonous seeds. Many Australian native plants have toxic parts (Covacevich et al., 1987), although the fruits of most are not palatable, nor attractive in appearance; an exception is the finger cherry *(Rhodomyrtus macrocarpa)*, a small rainforest tree from Queensland which is sometimes grown as an ornamental plant in gardens — the attractive fruits contain a toxin which causes blindness if eaten. There is also the burrawang *(Macrozamia communis)*, the large, handsome seeds of which are quite poisonous unless crushed and left in moving water for a few days.

Intense irritations or allergic reactions to plant products have long been recognised, such as skin irritations from contact with various members of genus *Rhus* (poison ivy, poison oak, poison elder, grown for their ornamental value in gardens), which produce an oil called uroshiol which has the peculiar property of creating no irritation on first contact, but of subsequently sensitising people to varying degrees, so that subsequent contact may be quite traumatic. There are also common nettles *(Urtica urens* and *U. incisa)*, unwanted introductions to Australia, which possess no serious toxin, but do have stiff, brittle hairs on the stems and leaves which readily penetrate the skin and cause intense irritation for a while. In the Australian rainforests, bushwalkers soon learn to avoid contact with the stinging tree *(Dendrocnide excelsa)*, a giant relative of the nettle, the leaves of which may be a third of a metre across, and covered with thousands of rigid stinging hairs. There is also the Gympie stinging tree *(Dendrocnide moroides)*, actually a shrub with large, somewhat heart-shaped leaves. Contact with the hairs causes swelling of the lymph glands and great pain which may persist for months.

Airborne plant materials, especially certain pollens, have been identified as important factors in promoting onset of rhinitis (hay fever) or asthma. Pollinosis is generally recognised as a condition which has links to asthma worldwide (Charpin et al., 1988), with the most widespread pollen allergy being grass pollen allergy, especially to ryegrass (*Lolium perenne*), Kentucky bluegrass (*Poa pratensis*), couch (*Cynodon dactylon*), and canary grass, *Phalaris aquatica* (Pollert, 1988; Reid et al., 1986; Bass and Baldo, 1990; Singh et al., 1992). Tree pollen, especially from birch (*Betula* spp.) and white cypress pine (*Callitris glaucophylla*) is recognised as a causal factor in rhinitis and asthma (Odei et al., 1986; Hemmens et al., 1988a, 1988b; Bass, 1991), with privet (*Ligustrum vulgare*) various *Casuarina* spp., olive (*Olea europa*), and, in New Zealand, *Pinus* spp. also implicated (Bass, pers. comm.; Fountain and Cornford, 1991). The wattles (*Acacia* spp.) have not been adequately studied, but few people seem to be affected (Bass, pers. comm.). The most significant weeds are plantain (*Plantago lanceolata*), the ragweeds (*Ambrosia* spp.), pellitory (*Parietaria judaica*), and Paterson's curse, or Salvation Jane (*Echium* spp.). Allergic people have an over-reactive immune system, so that, in addition to the normal function of the immune system in providing protection against infection, it causes inflammation and allergic symptoms. The immediate cause of the allergic symptoms is over-production of a specific antibody, Immunoglobulin, or IgE, in response to one or more *allergens*, normally harmless substances such as the pollens noted above. The IgE antibody binds to *mast cells*, found mainly in the skin and lining tissues of the nose and lungs. Inhalation of the allergen then stimulates the mast cells to release substances which cause contraction of the muscles surrounding airway passages, swelling and irritation of the tissues, and secretion of excessive amounts of mucous.

Hazards in the form of unwanted plants or weeds, such as prickly pear, Paterson's curse, bitou, or water hyacinth in Australia, indirectly affect humans, as weed infestations can significantly reduce the yield of valued plants, introduce toxic substances into human or animal foodstuffs, degrade valued ecosystems, and even create navigation hazards. Prickly pear (*Opuntia stricta*), spread rapidly in Queensland and northern New South Wales following its introduction, possibly as a garden escape, in the late nineteenth century, and by the 1930s it had covered 24 million hectares. As there are no native cacti, nor cacti grown for economically important reasons in Australia, spectacular success in biological control was achieved by introducing a cactus specific worm, *Cactoblastis cacto-*

rum, from Argentina. More persistent, in southeast Australia, is Paterson's curse, or Salvation Jane (*Echium* spp., of which *E. plantagineum* is most widespread), widely recognisable in pasture lands for its bright purple flowers. However, its 'curse' lies in that it smothers pasture, and then leaves bare ground when it dries out.

Bitou (*Chrysanthemoides monilifera* ssp. *rotundata*) was inadvertently introduced to New South Wales from southern Africa (where it is endemic) over one hundred years ago. Bitou possesses a number of attributes which enable it to compete successfully with and, in many cases, to displace, native vegetation in Australia, notably: prolific seed production; freedom from its natural diseases and predators; extensive tap and lateral root systems; rapid growth; adaptations to moisture stress; year-round flowering and seeding; and high rates of seed longevity and viability. Although bitou bush is presently preserving extensive areas of coastal sand from the hazard of sand drift, its tendency to develop into mono-specific stands at the expense of native species is detrimental to long-term coastal dune stability, and threatens both the aesthetic and biological heritage of the coastline. Pursuit of a biological control agent is under way, and some trial releases of candidate insects have been made, but it is too early to evaluate results.

Water hyacinth (*Eichornia crassipes*), a floating perennial plant native to tropical America, has widely been introduced elsewhere in the world, in most cases for its attractive mauve coloured blossoms, which also bring it the name 'water orchid'. The fleshy floating leaves are supported by bulbous, inflated petioles with internal intercellular air spaces, and the long roots, pendant from thick horizontal rhizomes, may hang freely in deep water, or, in shallow water, penetrate the soil beneath. Water hyacinth has escaped from cultivation and become a weed in many warm temperate and tropical waterways, where the impenetrable masses of tightly bound plants present a serious hazard to navigation, as well as disrupting freshwater ecology.

MICROFAUNA

Many tiny animals pose a threat to human life, health, or comfort. The principal impact of mites or lice may be to annoy, but others, such as the internal parasites of humans which cause schistosomiasis or African sleeping sickness, are significant hazards. Microfaunal hazards may also indirectly affect humans, in the form of parasites of livestock on which we depend for food or fibre, for example, ticks or internal parasites of cattle.

Schistosomiasis (sometimes called bilharzia) is one of the most serious human parasitic infections. It is caused by blood flukes (tiny flattened worms, leaflike in appearance) of the genus *Schistosoma*, and affects millions of people in tropical Africa and Asia, and in the West Indies. The fluke requires a water snail host for part of its life cycle, and the larvae enter the bodies of people through contact with infected snails or water contaminated with the larvae. Once in the circulatory system, the parasites usually concentrate in the pelvic area (particularly liver and spleen) and intestines. Fatalities directly due to schistosomiasis are not common, but the severe debilitation and malaise, and reduced life expectancy associated with the disease, along with the difficulty of cure, render it a significant hazard to people in regions where the parasite is endemic. African tsetse fly, *Glossina palpalis*, is the principal carrier of the parasite *Trypanosoma brucei*, which is the causative agent in African sleeping sickness, a debilitating and often fatal disease which has resulted in morbidity and mortality for millions of African people.

Mosquitoes are not only a nuisance, but a public health hazard as well, being carriers of disease to humans and other animals. The role of mosquitoes (*Anopheles* spp.) in the transmission of malaria is well known, but in Australia, Murray Valley encephalitis virus, Ross River virus, dengue virus, and Kunjin viruses are all spread by mosquitoes. Significantly, some species found in Australia are capable of transmitting serious diseases (malaria, filariasis, yellow fever) which are not presently a problem in this country (Dale, 1993). Apart from hazard modification (insect screening of homes), control by chemical means (individual repellents or environmental pesticides), habitat modification, or introduction of a predatory fish (*Gambusia affinis*) have been the principal management techniques (Dale, 1992). However, there is now attention being given to avoiding the hazard by appropriate location of human settlements in relation to major mosquito-breeding areas (Dale, 1993).

MACROFAUNA

The folklore of most countries is replete with stories of encounters with the large, ferocious carnivores which presented a very real hazard to our ancestors who dared venture outside the relative safety of the village and its immediate environs, into the mysterious forest, the steppe, or the awesome mountain passes. Many of the tales survive to the present, and indeed, the writer can remember, as a very small child, feeling somewhat cheated that wolves and bears, the stuff of spine-chilling nursery story adventure, were not to be found in the Australian bush surrounding his parental home!

Unique among the inhabited continents, Australia boasts no large terrestrial carnivores. Bears and wolves in North America, tigers in Bengal, pythons in southeast Asia, and lions in Africa are but some of the creatures hazardous to humans. Many of these animals have suffered drastic decreases in numbers within the last century, due to the combined effects of hunting and habitat destruction, but there are still close encounters. There is conflict between those who wish to conserve the Bengal tiger, and the local human cohabitants of its range, who see it as a hazard, for example, and, at the time of preparing part of this book, the writer was in Denver, USA, when a jogger from one of the peri-urban ravine suburbs at the base of the Rockies was taken by a mountain lion.

Australia may be free of hazards posed by large, warm-blooded carnivorous animals, but there are some large and many small cold-blooded creatures which are capable of causing severe trauma, or even death, to human beings (the exception, perhaps not surprisingly in view of its oddity in almost all other respects, is the platypus, *Ornithorhynchus anatinus*, which, although a warm-blooded animal, boasts poison spurs on the hind legs of the male). Many people will never encounter them, but there is a surprising variety to be found in the Australian continent and coastal waters: venomous snakes and sea snakes, insects, spiders, ticks, centipedes, scorpions, various jellyfish and cone-shell fish, stinging fish, stingrays, the blue-ringed octopus, and, of course, crocodiles and sharks.

The sea wasp or box jellyfish (*Chironex fleckeri*) found in Australian waters ranks as one of the most venomous sea creatures and poses a hazard in tropical waters significant enough to prompt many experienced people to wear body suits of light material while swimming. *Chironex* has been implicated in the deaths of about sixty people: its 3 m long tentacles release a toxin which almost immediately affects heart function and breathing, and attacks red blood cells. Blue-ringed octopi (*Hapalochlaena lunulata* and *H. maculosa*) are tiny, hardly large enough to span the length of a hand, and readily identified by their blue spots or rings, especially if aroused. The few recorded fatalities have all occurred as a result of handling them.

In the interaction between humans and sharks in Australian waters, it is undoubtedly the shark which has been the loser. Over many years, widespread netting of sharks off popular beaches, and outright hunting, have reduced their numbers by many times the 180 fatal attacks on humans. When a fatality does occur, it sometimes initiates a shark pogrom. Any large animal, whether on land or in the sea, must be regarded as potentially dangerous, and there is no reason to suppose that

Table 1.1 Shark attacks in Australia, 1788–1993

State	No. of attacks	No. fatal	Latest
NSW	189	71	1993, Byron Bay
Qld	187	70	1992, Moreton Island
Vic.	22	7	1977, Mornington Peninsula
SA	29	15	1991, Aldinga Beach
WA	37	7	1967, Jurien Bay
Tas.	17	9	1993, South Cape Bay
NT	10	3	1938, Bathurst Island
Total	491	182	

Source: Australian Shark Attack File, courtesy John West, Taronga Zoo, Mosman.

animals such as the white pointer shark (*Carcharodon carcharias*), tiger shark (*Galeocerdo cuvier*), or various whaler sharks (*Carcharhinus* spp.), which have been identified in fatal shark attacks on humans in Australia, are especially malevolent. Indeed, these powerful animals are capable of inflicting much more damage on humans than they generally do, and the majority of people are bitten and released (J. West, Taronga Zoo, pers. comm.). It is probable that many shark 'attacks' are provoked as much by curiosity on the part of the shark, or by invasion of the shark's 'territory' by the human, as they are by feeding behaviour of the shark. Considering the huge number of person-hours spent in water contact recreation in Australian coastal waters, and the fact that the average rate of fatalities from shark attack is less than one per year (table 1.1), it may be observed that the drive to the site may well be the most dangerous part of a beach outing.

Popular works with titles like *Crocodiles: Killers in the Wild* (Hermes, 1987), and spectacular media accounts of the rare crocodile attack on a person, perpetuate the notion of the crocodile as a serious environmental hazard of Australian tropical areas. In fact, habitat modification, and hunting of crocodiles for their skins has eliminated them from their range in many parts of the world. Two species occur in Australia: the freshwater, or Johnstone's River Crocodile (*Crocodylus johnstoni*), which is endemic to northern Australia, and the saltwater, or estuarine crocodile (*Crocodylus porosus*), which has a natural range from Sri Lanka, around the top of the Bay of Bengal, and down the Indo-China peninsula through northern Australia, Papua–New Guinea and the Solomon Islands. Both are listed in the IUCN Red Data Book — the freshwater crocodile as 'vulnerable', and the saltwater crocodile as 'endangered'.

The saltwater crocodile can grow to enormous size — individuals to 9 m in length have been reported (Webb and Manolis, 1989), and there is photographic and documentary evidence of many specimens over 7 m, with mass approaching 1000 kg. Although protected by law from hunting over most parts of its range, the IUCN Red Data Book points out that, apart from in Australia, protection is largely ineffective. The freshwater crocodile is small (2–3 m or less) and has never been known to kill anyone, although bites result in trauma, and produce wounds which usually refuse to heal for long periods.

Along with the Nile crocodile (C. niloticus) in the old world, the saltwater crocodile, which has been observed at sea and on land far from its normal estuarine habitat, has been known to fatally attack and even to eat human beings in Australia. Perhaps the best documented attack concerns a pig hunter who was taken by a crocodile in the Northern Territory in 1975: a large crocodile subsequently shot nearby yielded the body of the hunter, in eleven pieces. Although it is certain that there were unreported attacks in the era when tropical Australia was a wild frontier, the number of fatalities remains small. The first recorded fatality (mid-nineteenth century) was of a seaman asleep in a small boat in the Roper River estuary, with his leg hanging over the side of the boat — a crocodile took the leg and the man followed. Subsequently, there have been less than twenty authenticated fatalities in Australia from crocodile attack. Most were of people swimming in waters known to be inhabited by saltwater crocodiles, and, as pointed out by Webb and Manolis (1989), the attackers were very large (>5 m) crocodiles which clearly were old enough to be survivors from the hunting era, and to have known harassment from man.

The most venomous (and potentially most dangerous) snakes in the world occur in Australia, and although there are on average only a couple of fatalities per year, the scientific, as well as popular literature, abounds with accounts of fatal or near-fatal snake bites. Every person in Australia exposed to the 'bush', whether occasionally, as a bushwalker, or more permanently, as a resident, has learned a healthy respect for creatures such as the death adder (Acanthopis antarticus), the taipans (Oxyuranus scutellatus scutellatus and O. microlepidotus), the tiger snakes (Notechis scutatus, N. ater ater, N. ater occidentalis, and N. ater humphreysi), and the king or the common brown snakes (Pseudechis australis, Pseudonaja textilis textilis), to name only the best known and probably most dangerous.

Prior to production of antivenom, 50 per cent of death adder bites were lethal. The small (usually <0.6 m) grey to greyish-red snake is characterised by a broad triangular head, stubby body, and thin tail. It will not usually retreat if approached by humans. Once widespread in Australia, the death adder has virtually been eliminated over much of its range by habitat modification and predation by feral animals.

Based on a rating system for venom yield and toxicity, fang length, temperature and frequency of bite, the Queensland Museum rates the taipan as the most dangerous snake in Australia (Mirtschin and Davis, 1982). *O. scutellatus* is found in moister coastal areas around northern Australia and as far south as far northeastern New South Wales. It is pale creamish around the head, with brown, coppery-red or even olive or black on the body, can grow over 2 m long, and is nervous and alert, with acute sense of smell and vision. Consequently, it is difficult to surprise, and will usually retreat from humans. The rare fatal bites have been inflicted by snakes approached suddenly or cornered. Introduced rats (*Rattus rattus*) have provided an attractive food source. *O. microlepidotus* is found in the Channel Country and environs, where its ecology is thought to be closely associated with the occurrence of the plague rat (*Rattus villosissimus*), on which it feeds (Mirtschin and Davis, 1982). It is pale to dark brown above, with dark flecks, yellow below, and has a dark brown to black head.

There are many hazardous insects and arachnids (the spiders, scorpions, ticks, etc.), and each country has its particularly dangerous representatives — tsetse flies in Africa, deadly *Centruroides* scorpions in Mexico, and black widow spiders in the Americas, for example. In Australia, there are, on average, 2–3 deaths from bee stings each year, and few deaths in all from bites and stings of venomous animals, insects and spiders (figure 1.1), but no insect or spider is feared as much as the funnel-web spider (members of genera *Atrax* and *Hadronyche*). Funnel-web spiders (figure 1.2) are large and black, with a shiny, hairless head, closely grouped eyes, and spinnerets that project noticeably past the body. They are frequently large enough for the legs to span the width of a man's hand. Some bites do not result in envenomation, but the length of fang and force of the strike may produce trauma which is exacerbated by the acidity of the venom (Gray, 1992). Severe local pain, erection of hair, sweating, and redness due to capillary congestion usually follow the bite, and, in severe cases, disorientation, severe vasoconstriction, abnormally low blood pressure, deep coma, and cardiac arrest may occur (Austin and Heather, 1988). A funnel-web antivenom was developed in 1980, but the most effective management strategy consists of alertness and due caution!

Through their impact on crops and stored foodstuffs many creatures are hazardous to human well-being in that they compete with humans for available nourishment. But it is when these creatures appear in plague proportions that a serious natural hazard is presented. Although Australia has its share of insect hazards, such as locusts, mammal species

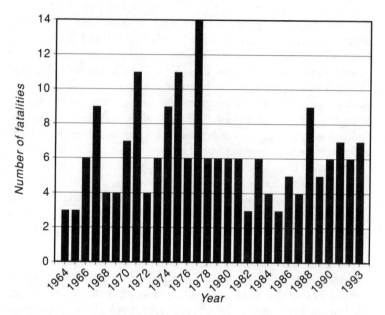

Figure 1.1 Fatalities in Australia from bites and stings of venomous animals or insects. Includes spiders, snakes, bees and other insects, as well as venomous sea animals.
Data source: Australian Bureau of Statistics.

giving rise to plagues or environmental hazards have largely been imported. Twenty-six species of wild mammals (including twelve domestic species which escaped and established feral populations) have been introduced to Australia in the last 200 years (Wilson, 1992). Many of these, such as feral pigs, goats, buffaloes, and rabbits have become serious environmental hazards, and (especially in the case of the rabbit) serious threats to human well-being as competitors for the plant products used either directly or indirectly by us. Mice sometimes break out in plague proportions, especially in South Australia and western Victoria, and there is also the indirect effect on humans of rodents which act as vectors for various diseases such as poliomyelitis or favus ('ringworm' — actually a fungus): Hadlington and Gerozisis (1985: 246) list twelve diseases which may be transmitted from rodents to people in Australia.

The decision to introduce the rabbit (*Oryctolagus cuniculus*) into Australia in 1859 resulted in what may rank as the single worst environmental disaster of all time. For a period at least, the rabbit was probably the quantitatively dominant mammal in the continent. Its effect as an

Figure 1.2 When disturbed, a funnel-web spider quickly rises to a striking position, with the front two pairs of legs off the ground, and a drop of venom may appear at the tips of the fangs.
Atrax robustus (♀), photograph courtesy Dr M. Gray, Australian Museum.

environmental hazard, destroying plant cover and displacing both native and introduced herbivores, and outbreak in plague proportions prior to biological control, has become part of Australian folklore. Some idea of the magnitude of impact may be gained from the fact that, in the eight years subsequent to its arrival in New South Wales alone, while it was still considered feasible to control numbers by hunting, 270 000 000 scalps were turned in to government inspectors in exchange for bounties. Biological control by means of myxomatosis virus and the rabbit flea now helps suppress numbers, and there have been no outbreaks of plague proportions since myxomatosis became established.

Both grasshoppers and locusts belong to the family *Acrididae*, and have been known as serious biological hazards from ancient times:

The locusts . . . invaded all Egypt and settled down in every area of the country in great numbers. Never before had there been such a plague of locusts, nor will there ever be again. They covered all the ground until it was black. They devoured . . . everything growing in the fields and the fruit on the trees. Nothing green remained on tree or plant in all the land of Egypt. (*Exodus 10:15* (about 1500 BC))

There is also an interesting bas-relief from eighth century BC Nineveh in the British Museum which depicts what might be called 'locust kebabs', and suggests they were in turn eaten by humans!

Locusts exist in most parts of the world, but their impact is particularly devastating in marginal semi-arid zones that border on the deserts. Locusts are present in small numbers at all times. Their numbers are held in check — as it is normally true for all organisms — by predators and by diseases. The coincidence of drought with population explosions of this insect has resulted in some of the worst famines the world has known.

Both adult and nymphal stages of the insects are general defoliators: although they show a preference for grasses (*Gramineae*) as food, they are not restricted to them. In places where natural grasslands are used for pasture, or have been ploughed for cereal growing, *Acrididae* are often the dominant insect fauna, and may be serious pests. Locust populations periodically explode and pose a dual threat because the plague seems to occur most effectively when drought also occurs. A locust consumes its own weight in food every day, and the collective destruction caused by a large swarm, with a total mass of tens of thousands of tonnes, can be total.

True locusts are dimorphous (occurring under two distinct forms), with solitary and gregarious phases that differ in both colour and behaviour, with the gregarious phase only being migratory. There are a number of principal plague forming species: *Chortoicetes terminifera* (the Australian plague locust), *Dociostaurus maroccanus* (Mediterranean), *Locusta m. migratoria*, *L. m. migratorioides*, *L. m. maniliensis*, *Nomadacris septemfasciata*, *Patanga succincta*, and *Schistocerca gregaria*, the desert locust (Asia and Africa).

Locust eggs are laid in pods in the soil. Egg pods may contain 20–80 eggs, with each female laying several pods. Hatching is facilitated when the soil is warm, sandy, and endowed with some moisture. If high moisture conditions prevail, however, the eggs of the locust may rot. Sandy soil not only facilitates the placing of the eggs, but also enables the locust nymphs to reach the surface easily after hatching. There is usually only one generation per year, but up to three in plague years. Hatching may occur over many square kilometres, at densities exceeding 1000 nymphs per square metre. The newly hatched locusts are wingless and move along the surface by hopping, in search of food. Their wings develop after five stages of moulting which takes about five weeks.

When excessive numbers of young locusts survive, they quickly destroy their local food supply; they then migrate toward regions which offer more vegetation. Plague outbreaks may last for several years and are characterised by many swarms of adults and of 'hopper' bands, the latter being a cohesive mass of nymphs which increases in size and speed of movement during the five instars (inter-moult stages). In Africa, mature locusts have been known to migrate from the Sahel across the Sahara and as far northward as the coast of the Mediterranean, where irrigated crop areas are exposed to the ravaging insects. A plague which started in the Sahel in 1986–87 reached northwest Africa at the end of 1987. It expanded in 1988 in north Africa, the Sahel, the Sudan, the Near East and southwest Asia, and in October 1988 swarms crossed the Atlantic to Caribbean countries (Skaf et al., 1990). Plagues are separated by recessions when gregariously behaving collections of locusts are few, small, and transitory, and may be completely absent. Swarms are displaced downwind, and may move 100 km or more in a day. They may be seen from an aircraft several tens of kilometres away in clear weather.

Given the biology of the insect which allows an extremely rapid rate of multiplication when environmental conditions are suitable, the most important factors, at least with respect to the Australian plague locust, are meteorological. Rain is the most important, as it promotes the vegetation growth necessary for locust development and lipid accumulation necessary for adult migration, will restart development of eggs in quiescence after dry conditions, and allow any adults present to mature and lay. Warm temperatures promote rapid development at all stages of the life cycle, and the strength, timing, and direction of winds will determine if, when, and where locusts will migrate (Bryceson and Wright, 1986).

There are essentially three possible control strategies (Symmons, 1992): to prevent the plague by control during the upsurge stage; to eliminate the plague by destroying nearly all the locusts; and merely to protect crops and allow the plague to follow its natural course.

Pesticides have been applied by airplanes, by truck-mounted sprayers, and, along with poisoned baits, by thousands of local workers. While these efforts have been called successful, it must be recognised that there is political pressure on those in charge to claim a success, and the poisoning of the environment has increasingly created concern. Symmons (1992) pointed out that the application of millions of litres of pesticide, at a cost of about half-a-billion dollars, during the 1986–89 desert locust plague in northern Africa, had a doubtful effect and concluded that plague prevention is probably not technically feasible. Restrictions on the use of persistent pesticides such as dieldrin have resulted from environmental concerns, but created major technical, logistic, and financial problems, as use of non-persistent pesticides require the finding and spraying of individual bands (Skaf et al., 1990). Air-to-air spraying, to destroy swarms in flight, appears best, with perhaps a couple of million litres of ULV (ultra-low volume) pesticide required to control a swarm of 10^{12} insects: the air-to-air technique also minimises environmental damage, as there is but a small deposit per unit area.

Other methods less damaging to the environment have been recommended: early-sown crops tolerate damage much better than late-sown crops because early crops make considerable growth before the grasshoppers hatch. Grain crops which have developed seed heads can withstand quite extensive defoliation without serious reduction in yields, and in agricultural areas ploughing moves the eggs to the surface, where they are picked up by birds or are destroyed by exposure to sun and wind. Controlled flooding of fields can also be used to kill off the eggs as well as newly hatched nymphs. A pest is of course only an economic pest above a certain population density, and control measures are usually designed only to lower the population below the density at which it is considered an economic pest: only rarely is complete eradication attempted.

CONCLUSION

In the technological era, our interactions with biological competitors in the struggle for existence have been marked by the language and tactics of war — nothing less than extermination was the aim. For some of these biohazards, such as smallpox anywhere, or rabbits in Australia, that was entirely appropriate. For others, our increasing ecological knowledge is enabling us to realise that sustainability of human existence on the planet demands that we learn to co-exist with organisms that our grandparents may have dismissed as 'pests'. The conflict between beach user and shark in Australia, for example, has resulted in the death of about 400 sharks for every human, and if one includes commercial hunting of sharks in Australian waters, the ratio is about 150 000:1. Unlike 'nice' creatures such as whales and dolphins, sharks command little public sympathy, but are they less important to the sustainability of the balance of life on this planet? The example of the shark may seem extreme, and probably is, but it underscores the situation of many organisms on which we have waged war, at enormous economic cost, and emerging environmental cost in terms of persistent biocide pollution and ecosystem simplification. Newer approaches which emphasise pest-resistant crops, and biological controls, including measures such as the use of genetically modified viruses to deliver immunological sterilising agents to target animal species, promise to give us the capacity to 'fine tune' ecosystems, with potentially more harmonious relations between humans and other organisms. But many of these new techniques are not without their own dangers. Hazard avoidance, allowing at least some of these organisms their own space, an approach which is now beginning to be adopted in some areas for mosquito management, for example, may well be the least costly solution in the long run. 'The flower must tolerate a few caterpillars if it is to be acquainted with the butterflies.' (Antoine de Saint Exupéry, *The Little Prince*).

2

Bushfire or wildfire — Apocalypse in our time

Fire has long been a part of the Australian scene, and the accounts of many early explorers are liberally sprinkled with comments on the number of fires which they observed burning in the landscape. Most Australian vegetation types are fire adapted (Gill, 1975), and the dominance of sclerophyll eucalypt forests, and consequent high fire hazard (Cunningham, 1984), in much of southeastern Australia is associated with frequency of fire over many thousands of years. Fire has played a prominent role in shaping the Australian biota, and human societies, beginning with the Aborigines and continuing with the European settlers, have used and misused fire in ways that have reshaped the environment into something more suitable for their purposes. Pyne (1991) shows that geography, climate and vegetation (especially the pyrophilic eucalypts) shaped and were shaped by fire, and explains that the early Aborigines adapted to this fire-dominated land and, with their ubiquitous firesticks, central to their lifestyle as a hunting and foraging people, set in motion processes which created a complex landscape.

In the forested zone, and especially near urban areas, frequency of fire has increased following European settlement, but in many of the grassland areas, frequency of fire has fallen. Use of fire as a tool for clearing the land, and subsequently in some aspects of land management, has been widespread. Fire was used to clear forest, fire has been used in grazing areas in the belief that a burning over of dry and dead vegetation in the wintertime encourages a fresh flush of green grass suitable for cattle grazing in springtime, and fire has been used in an attempt to reduce the hazard of bushfire by lighting 'cool', slow burning fires on still days during the coldest part of the year to consume accumulated litter in forests, particularly those surrounding urban areas. Proliferation of large-scale burning to reduce fuel loads has led to conflict between advocates of organised fire management and environmentalists who see these 'prescribed fires' as an unnatural and ecological disaster. There is an uneasy truce between fire managers wanting to reduce fuel loads and the increasingly urban population willing to take the risk of devastating fire as a price for enjoyment of 'wild' areas. However, a large area of forest which has not burnt for some time contains an enormous amount of stored solar energy, which can be released in a few hours by bushfire, resulting in energy emission equivalent to that of a nuclear bomb (cf. figure 1, p. 4).

The majority of the Australian population resides in urban centres and has had little experience of fire, even on the margins of the forested zones. A potentially serious fire season does not arise every year. Consequently, the general awareness of fire danger and of the risk from foolish action with fire is low. People are more aware in the rural grassland areas, where there is a regular high-risk period every summer and they have more evidence of the result of the accidental release of fire. Mediterranean climates, with hot dry summers, as are experienced in South and Western Australia, and California, pose an especially high fire risk.

Wildfire is strongly dependent on environmental variables, as well as on available fuel (figure 2.1), but in Australia, as in North America (Todd and Kourtz, 1991), human interference is crucial with regard to ignition (table 2.1). A thirty-year study by Flannigan and Harrington (1988) in Canada showed that long sequences of days with less than 1.5 mm of rain or of days with relative humidity below 60 per cent were highly correlated with area burned, but meteorological variables explained only one-third of the variance. Simard and Main (1987) found that the variance of wildland fire activity was double that of the underlying causative factors (weather and fuel), suggesting that fire outbreak prediction will remain an elusive goal. Nature may provide the fuel and fire weather, but deliberate, unintentional, or unthinking misuse of fire by people is the principal factor.

THE PHENOMENON OF WILDFIRE

A spreading forest fire is a complex combustion process in which the flaming front first heats, and then ignites, unburnt woody and herbaceous fuels. The modes of heat transfer responsible for fire spread from the flaming front are principally by convection and radiation. Conduction does

Table 2.1 Principal causes of bushfires

Cause	% of fires	Cause	% of fires
Burning off — legal	12.3	Campers	1.7
Burning off — illegal	15.3	Domestic, children	4.3
Lightning	5.6	Industry	0.3
Power lines	1.8	Smokers	1.5
Rubbish tips	3.2	Arson	8.4
Sawmills	0.3	Miscellaneous, known	9.7
Transportation	8.1	Miscellaneous, unknown	27.5

Source: Average of data held by the NSW Dept. of Bushfire Services.

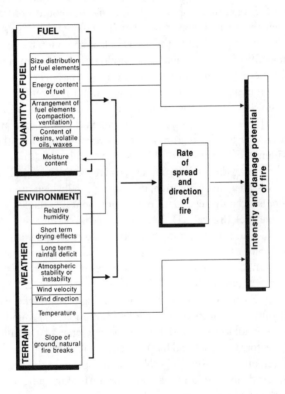

Figure 2.1 Factors contributing to bushfire.

not contribute significantly to fire spread because wood and soil are poor heat conductors. In the heating process, the moisture in the fuel is first evaporated (temperature $\geq 100°C$); the terpenes, fats, oils and waxes are vapourised ($100°–200°C$); the cellulose is thermally broken down and its breakdown products converted into vapour (temp. $> 200°C$); and finally the volatiles ignite to form a visible flame (300°–400°C). Some plants, particularly members of *Eucalyptus* and *Pinus* genera, carry significant volatile waxes and oils which release high heat per unit mass and which also increase the flammability of the plant fuel. The lignins (cross-linked phenolic polymers which combine with cellulose to give woody plant tissue its rigidity) are very stable when heated and produce a carbonaceous compound which largely remains in the char. After the flaming combustion has ignited and burned most of the volatiles, the remaining carbonaceous material may burn as a solid by surface oxidation called glowing combustion. Flaming and glowing combustion are not discrete events in forest fires because of the complex mixture of fuel sizes, moisture contents and arrangements. However, the flaming front is dominated by combustion of gases and glowing combustion occurs primarily after it passes.

WILDFIRE FUEL

Fuel size

Ignition is facilitated by a large area to volume ratio of fuel elements. Consequently, the small fuel elements such as grass, leaves, bark and twigs constitute the most flammable proportion of the fuel available to a bushfire. Large fuel elements such as tree trunks, branches and fallen logs do not constitute a readily available source of fuel, and indeed may be seen smouldering for days following the passage of a bushfire.

Fuel arrangement

In addition to the size of fuel elements, their arrangement in three-dimensional space greatly affects the supply of oxygen and therefore the ease with which the fuel may be burnt. Very compact fuels such as peat will burn quite slowly, even when dry, compared, for example, to a similar amount of combustible material in the form of dead grass or loosely packed forest floor litter. In the dry sclerophyll forests of southeastern Australia, fuel arrangement for combustion is near optimal when fuel accumulation has reached an equilibrium with fuel decay. In addition to the fine fuel elements such as dead leaves and twigs, an additional source of fuel in this type of forest is the long shards or strings of bark which frequently festoon the branches and trunks of trees and aid the spread of fire, since they may be lit from beneath by a fire which is otherwise restricted to the undergrowth. Unmanaged pine forests with dead lower branches, heavy accumulation of dead needles, and resinous green needles are also near optimal in their arrangement of fuel elements.

Fuel quantity

The quantity of available fuel is commonly expressed as mass per unit area, usually tonnes per hectare. Concepts which relate the mass of fuel to its enclosing volume, in an effort to express both the availability of the fuel and its arrangement, as discussed above, have been experimented with, but no effective and standardised method has been reached. Not all potential fuel may be available. For example, total biomass above the ground in a dense forest may be greater than 500 tonnes per hectare, and even in a dry sclerophyll forest may exceed 250 tonnes per hectare, but of this, only 10–20 per cent may be immediately available for consumption by wildfire.

The gross accumulation rates of forest fuels are generally in the order of 1–10 tonnes per hectare per year, but net accumulation depends on the balance between the production of new fuel elements (or new biomass), which is greatly dependent on weather and climate, and the decomposition of litter, which is dependent on the population of microfauna and fungi as well as the weather and climate (and in particular the microclimate of the litter bed). Following a fire, there is usually a period in which the rate of litter accumulation exceeds the rate of decay, until equilibrium between decay and production is achieved. In wet sclerophyll forests, the period may be five to ten years following a fire, while in dry sclerophyll forest (in which up to 40 tonnes per hectare of fine fuel may accumulate), the equilibrium may not be achieved for as many as 25–30 years.

Fuel moisture content

The moisture content of fuel varies in different parts of a growing plant but, in addition, the dry fuel or litter will vary in moisture content according to the weather and the moisture content of the substrate. Since dead plant material absorbs moisture from the air, there is a relationship between the relative humidity of the atmosphere and the moisture content of fuel. Relative humidity is a function of weather conditions, but also varies on a daily basis, and is at its lowest during the warmest part of the day, usually mid-afternoon, and at its highest in the early hours of the morning. When the moisture content of forest fuel is above 15 per cent, fire intensity is low, and the fire is usually controllable; 20 per cent is the approximate upper limit for combustion (fires often die out upon encountering fuel of high moisture status). Combustion rate increases rapidly with decreasing moisture content, and when below 7 per cent it is critical for fire behaviour — at 5 per cent the rate of fire spread is double that at fuel moisture content of 7 per cent, and when the fuel moisture drops from 5 per cent to 3 per cent, another doubling occurs.

Fuel energy content

The measure of heat released by forest fuel is called the heat of combustion, and is the maximum heat that could be released by a dry fuel if it were completely burnt. The heat of combustion of average forest fuel is of the order of 20 000 kilojoules per kilogram (a watt, the normal unit of electricity consumption, is equivalent to one joule per second). However, the moisture content of fuel, and the fact that combustion is not complete (all forest fires are characterised by smoke which is largely composed of small unburnt particles of fuel, and all fires leave behind charred, incompletely burnt fuel), affect the actual *yield* of heat from forest fuel. *Heat yield* of forest fuels is usually taken to be 16 000 kilojoules per kilogram on average (Luke and McArthur, 1986).

FIRE BEHAVIOUR

The *heat output* from burning forest fuel is, however, a function of the behaviour of the fire. The measure used for heat output is called *fire intensity*, and is the *rate* at which heat is given off by a fire. It is a more useful measure than is temperature. The temperature of a burning twig can be the same as that of a large crown fire, yet clearly the crown fire is transferring more heat from the flaming front to the immediate environment than is the burning twig. Heat transfer (intensity) causes the adjacent fuels to be heated and burn, thereby releasing more heat and propagating the fire. Death of and injury to plants is also dependent on the heat transferred to them and on how much of the heat is absorbed. The standard formula used to estimate fire intensity is:

$$I = HWR/600$$

where

I = intensity of thermal output, kilowatts per metre of fire front
H = heat yield of fuel, in kilojoules per kilogram
W = available fuel, in tonnes per hectare, and
R = rate of movement of the fire front, in metres per minute.

The variation of intensity (I) is most significantly influenced by the variation in the rate of movement of the fire front. As we have seen above, the heat yield (H) is for practical purposes a constant, and fuel availability varies by about ten-fold, while rate of spread (R) varies about 100-fold. A relatively slow moving fire (say 10 metres per minute) burning in a medium fuel load of 10 tonnes per hectare would thus produce thermal output of the order of 2700 kilowatts per metre. The threshold of controllability for forest fires is considered to be around 4000 kilowatts per metre, while a 'controlled burn' for fuel reduction is held to below 500 kilowatts per metre if possible. Forest fire intensities have been estimated to reach as high as 100 000 kilowatts per metre (Kiil and Grigel 1969, Alexander 1982), although 60 000 kilowatts per metre is a more likely extreme, which would be represented by a fast fire moving at over 3 kilo-metres per hour in a heavy fuel of around 40 tonnes per hectare.

The radiative output of heat from a forest fire is great and may be responsible for pre-heating and drying adjacent fuel elements, thereby preparing them for the fire. Convection, although mostly vertical, may be affected by wind and/or by the characteristics of the terrain in which the fire is burning. Consequently the pre-heating of higher layers of fuel, the transfer of fire to tree crowns and the heating and drying of fibrous bark elements may all be accomplished by radiation and/or convection from an existing fire.

Wind is one of the dominant factors affecting fire behaviour. Wind has the effect of driving flames into unburnt fuel and increasing pre-heating by radiation, but it also carries burning embers ahead of the flaming fire front, accelerating fire spread. The rate of spread of a fire front is mea-sured as the distance (not area) covered by the moving flame front per unit time, and is very largely a function of the square of the *effective* wind velocity. The latter is the wind velocity near the ground inside the forest, and is substantially less than that in the open, above the forest canopy. The effect of slope is also highly significant. Since on a steep slope fuel ahead of the fire is exposed to additional convective and radiant heat, by the time a slope of around 25° is reached, a fire is virtually a sheet of flame parallel to the ground surface itself. Compared to conditions on level ground, fire velocity is approximately doubled on a slope of 10° and quadrupled on a slope of 20°.

Atmospheric conditions also greatly affect the behaviour of a forest fire. Under neutral atmospheric stability, a fire will develop its own con-vective column, but in unstable air a strong convective column with positive feedback producing indraughts and updraughts will develop. Under conditions of atmospheric inversion (especially if the inversion layer is low) a fire may be inhibited, but if the inversion is punctured, a tornado-like firestorm may be set up.

The behaviour of a fire may be thought of in terms of three stages, which each reflect a particular set of conditions:

1   A fire burning *under environmental control* where the behaviour of the fire is dominated by environmental conditions and there is no develop-ment of a distinct convective column. Such a fire will usually burn slowly, within the litter layer on the forest floor, at an intensity of less than 500 kilowatts per metre.

2   The next phase is one in which there is an *interaction* between the fire and environmental conditions such that convective circulation is set up by the fire, with the depth of the flame large in relation to its width. The flame zone then becomes a source of heat and creates a feedback situation leading to the pre-heating of contiguous fuel elements, and the heat output causes a significant perturbation of the surface wind field.

3   The third condition is one in which the *fire is the dominant factor* in the environment. A well-developed, towering convective column of circulation is set up, producing a firestorm. *Brands* (burning fuel elements, particularly shards of bark) are carried aloft by the convective circulation, and then downwind by the ambient wind field, resulting in *spotting* (initiation of fires well ahead of the principal fire front). Spotting at distances of a kilometre is common in eucalypt forest fires, and has been observed up to 20 kilometres. It is a significant factor in the catastrophic potential of bushfires in Australia, creating new fire fronts well ahead of the source fire, and enabling bushfires to 'jump' obstacles, such as major rivers: fire-suppression crews may require hours to deploy at the new location(s).

The principal factors contributing to forest fire behaviour were first formalised by Australian forest researcher A.G. McArthur in the form of a circular slide rule (the 'McArthur meter', well known bushfire managers). The meteorological variables contributing to fire behaviour are integrated into a single index of fire danger on a 0–100 scale, which with a knowledge of fuel load, enables a fire manager to predict fire intensity (figure 2.2). At a value of 1, a fire might be expected to be extinguished by environmental variation, while a value of 100 on the McArthur scale corresponds to a 'worst-case scenario' for bushfire in Australia. As an example of the interaction of variables contributing to the McArthur index and their importance, if there was a condition of serious moisture deficiency for three months, with no rain at all for two weeks, and air temperature at 40°C (not uncommon conditions in the Australian summer), then the McArthur fire danger index for a range of typical wind velocities and relative humidities would be as shown in table 2.2. Fire hazard indices have also been developed for forest environments in Canada (van Wagner, 1987), France (Carrega, 1991), and the USA (Rothermel, 1983; Bradshaw et al., 1983).

IMPACTS OF BUSHFIRES

Impacts of bushfire on forest and peri-urban housing are dramatic and obvious, but wildfire causes problems for most land managers. Managers of pastoral country fear wildfire because it may sweep away precious feed and destroy both stock and infrastructure. Those responsible for

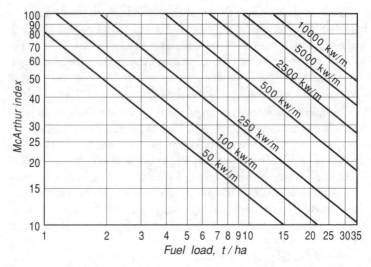

Figure 2.2 Fire intensity as a function of the McArthur fire danger index and fuel load in tall eucalypt forest

conservation of lands are charged with maintaining the full diversity of plants and animals, a task made difficult by wildfire, especially if reserves are small. In addition, there are the opposing pressures on park and forest managers to preserve their lands in as natural a condition as possible but also to reduce bushfire hazard by fuel reduction burning. Fire severity cannot be defined with a single measure; rather it is a concept. Units of measure must be specific to the system of interest. There are five systems of interest: ecosystems (flora and fauna), geosystems (soil and water), the atmosphere, fire management, and society. For ecosystems, fire severity could be measured in terms of disturbance characteristics (percent of plants and animals killed, effects on soil organisms, post-fire resprouting, etc.). For geosystems, measurements would focus on off-site movement of

Table 2.2 Typical values of fire danger

Based on McArthur's Fire Danger Meter and air temperature of 40°C, severe moisture deficiency, and no rain at all for two weeks

Relative humidity	Wind velocity, kilometres per hour							
	0	10	20	30	40	50	60	70
15%	25	36	45	56	70	95	100+	100+
30%	17	22	26	34	44	56	70	90
60%	6	8	10	12	16	20	24	32

material (loss of soil nutrients, increase in peak water runoff, soil erosion). For the atmosphere, severity would be measured in terms of fire emissions (quantities of particulates, toxic gases, greenhouse gases, etc.). Fire management would be concerned with the number of crews deployed, mobilisation logistics, suppression effort required, losses and damage to equipment, etc. From a societal perspective, severity would be measured in terms of social change (number of homes damaged, deaths and injuries, net resource value changes, and so on). Up to 30 per cent of the total area of spinifex grassland in Australia is burnt each year (Morton and Andrew, 1987 — see figure 2.3), and in 1974, 3.4 million hectares of the Barkly Tableland and western New South Wales were destroyed by fire. Foresters and peri-urban dwellers are conscious of the potentially huge economic losses, and sometimes loss of life, from bushfire (table 2.3). Bushfires of 1967, in Tasmania, when 62 people died and 1718 houses were destroyed, ranked as the worst Australian bushfire disaster in living memory, prior to Ash Wednesday in 1983. On that day (16 February 1983), towards the end of the disastrous Australian drought of 1982–83, and after four weeks of high fire danger, air temperatures soared to 40°C, relative humidity dropped as low as 20 per cent, winds reached up to 72 km/h, and half a million hectares of desiccated grassland and forest ignited in a conflagration which killed 76, left about 8000 homeless, and caused direct losses of about $200 million in Victoria and South Australia.

Summer 1993–94 was a period of severe moisture deficiency over the state of New South Wales. On 6 January 1994, temperatures were high, mostly over 40°C, with some country towns in the northeast of the state recording new record maxima. In addition, there were unusual conditions of strong (up to 60 km/h), exceptionally dry westerly winds which prevailed for the next four days. Fuel load in much of the bushland was high, and the hot dry wind reduced the moisture content of fuel to danger level. So strong was the desiccating effect of the wind that many sensitive plants growing in domestic gardens, such as ferns and rainforest species, shrivelled, browned, and died. Fire danger on the McArthur meter hovered at or near the maximum value of 100. Conditions were highly conducive to catastrophic fire, if ignition occurred. And it did.

There was no question of trying to suppress the fires in most bushland outbreaks — with fire intensities in excess of 10 or even 30 million watts per metre width of fire front, they were unstoppable. Over 600 000 hectares of natural forest were consumed in two days, some in the lower catchment of the Lane Cove River, only 10 km from the Sydney CBD. Arson was considered the cause of several fires, and some arrests were

Figure 2.3 Huge areas of central Australia, frequently thousands of square kilometres, are burnt by grassfire each year, resulting in a complex landscape mosaic resembling bold brush strokes. Location: Northern Territory, centre of picture approx. 131.5°E, 21.0°S. Scale as printed approx. 1:1 000 000.
Source: LANDSAT image reproduced courtesy ACRES, Canberra.

made, but the spotting behaviour of well-developed eucalypt forest fires was responsible for creating numerous new fronts, as in the southern suburbs of Jannali and Como in Sydney, where the fires 'jumped' into a hitherto apparently safe area and consumed 87 houses. At the height of the crisis, at least 150, and probably close to 200 wildfires were

Table 2.3 Impact of bushfires — New South Wales, 1982–92

Year	Number of fires	Area burnt (hectares)	Damages, $m (1993 dollars)
1982–83	7 899	447 400	104.1
1983–84	2 640	20 954	15.2
1984–85	6 294	2 720 000	86.3
1985–86	5 126	128 548	36.3
1986–87	10 158	457 679	90.1
1987–88	6 593	209 642	58.0
1988–89	5 197	119 942	38.6
1989–90	9 203	294 413	65.5
1990–91	9 090	768 400	113.4
1991–92	11 605	500 400	76.3
means	7 380	322 000	68.4

Source: Annual reports, NSW Dept. of Bushfire Services.

burning — the number is uncertain, as individual fires sometimes coalesce, reducing the number, but not the hazard. Emphasis in the fire management effort was on protection of life and property, by emergency firebreak construction, either mechanically or by back-burning, and a herculean suppression effort. Volunteer firefighters were airlifted from all parts of Australia, and some interstate crews drove their fire-engines non-stop the thousand plus kilometres to fires in New South Wales. Some 2000 regular firefighters and 16 000 trained volunteers were co-ordinated in a para-military operation which was remarkably effective. In all, 203 dwellings and other structures were completely destroyed, and about 150 seriously damaged (figure 2.4), along with a score of motor vehicles, and there were four fatalities associated with the fires. The damage to forests and wildlife was enormous, including losses of production forest, but the extent was unknown as this book went to print. Urban damages were estimated at $152 m, and another $15 m was spent on firefighting. Such losses are regrettable, but relatively small compared with the potential: over 25 000 people at risk were evacuated, and many thousands of homes at risk, which could not be moved, remained safe due to the fire management effort. The response demonstrated how an effective fire management strategy, supported by efficient ancillary services (the Bureau of Meteorology set up a forecasting unit in bushfire HQ for the duration of the crisis, for example), and most importantly, in the Australian situation, an army of willing, well-trained volunteers, can succeed in minimising loss of life and property in a bushfire crisis.

Dramatic and environmentally disastrous wildfires of unprecedented size and intensity have in recent years also been reported from Southeast Asia and elsewhere. In Sabah (E. Malaysia) some 950 000 ha of forest

Figure 2.4   Home at Kenthurst, New South Wales, gutted by bushfire.
Photograph © Mirror Australian Telegraph Publications. Used by permission.

were burnt in 1983, not including secondary forest and agricultural land, while in East Kalimantan (Indonesia), drought and associated fire affected 3 500 000 ha of forest land comprising 800 000 ha of primary forest, 1 400 000 ha of logged-over forest, 750 000 ha of secondary forest and shifting cultivation areas and 550 000 ha of peat swamps and swamp-forest (UNEP, 1991). Wildfires in 1987 in the Da Xing An Ling area of northeast China, the most important timber-producing area of the country, destroyed 870 000 ha of forest, killed 193 people, and left another

56 000 homeless (Kreimer and Munasinghe, 1991). Fires on such a scale have attracted much attention, yet there is a lack of systematic information on wildfires in the tropics and subtropics which makes evaluation of their impact difficult. It has been estimated that a total land area of between 630 and 690 million hectares per year is burnt, most in the tropics and subtropics (Goldammer, 1988; Malingreau and Tucker, 1988).

FIRE MANAGEMENT

Bushfire fighters commonly refer to what they call 'the fire triangle', pointing out that the three arms of this triangle, all of which are necessary for fire, consist of: (1) oxygen, (2) fuel and (3) a source of ignition. However, in a bushfire situation, oxygen is always abundant, and there is little hope of excluding it from a fire as one would, for example, in an industrial situation where foam is applied to the burning object in order to exclude the supply of oxygen. Bushfire management must therefore concentrate on the other two arms of the fire triangle, i.e., on minimising the possibility of unintentional ignition, and on reducing or eliminating fuel from the potential path of a fire. The latter is usually achieved by means of 'prescribed burning' in which fires are allowed to burn, or are deliberately set, in conditions of fuel, weather and topography such that the burning will meet the objectives of fuel reduction without damage to economic resources, and will be contained to the area of interest. Fire suppression efforts ('firefighting'), used alone, lengthen the interval between fires but result in more intense and more extensive fires in drier seasons when suppression fails. Prescribed burning, on the other hand, by promoting surface fires in moister seasons, shortens the interval between fires, but reduces fire intensity and confines fires to smaller areas. In general, it is considered that a 50 per cent reduction in fuel will result in a 50 per cent reduction in the rate of spread of a fire but a 75 per cent reduction in the intensity of that fire.

In the last few decades, the introduction of prescribed burning under defined weather and fuel conditions has gained wide usage in peri-urban bushland, production forests, and even in some national parks. It is perhaps the cheapest management tactic used in the manipulation of native vegetation today. It is also one of the most controversial. In the early years of European settlement in Australia and in North America, fire was used (often indiscriminately) as a tool, to aid clearing operations and to improve the palatability of herbage for stock. This period was followed by another during which fire was seen as an enemy to be eliminated. Then, in recent times, renewed emphasis on the use of fire in the management of the hazard of uncontrollable, high-intensity fires has aroused great concern among those interested in the long-term conservation of natural resources. To use fire as a management tool, the manager has to have an understanding of the behaviour of fire and of its effects. Fire behaviour depends on topographic, meteorological and fuel variables which are largely independent of the actual species composition of a forest, and is therefore predictable to a certain extent over a wide range of landscape types. However, we do not have detailed knowledge on the effects of fires on individual plant species, and especially on the floristic diversity of forest types. The use of control burning as a fire management tool has been criticised on the grounds that it seriously affects the species composition of forests, especially the shrub and herb layer and their associated fauna. In particular, shrub species which require an interval of several fire-free years to reach maturity before producing propagules may be completely eliminated if control burning is carried out at intervals so frequent that such species never have time to become sexually mature. There is also the factor of plant mineral nutrient loss from an area, either in the ash which is carried away by the wind, or in ash which is washed out by subsequent rain before the nutrients are reincorporated into the biomass. Repopulation by fauna is slow, and in the case of remnant populations, or populations already subject to severe predation by feral animals, some species may be made generally or locally extinct by a severe fire. Thus, on the one hand, managers of forestlands are expected by society to maintain them in as pristine a state as possible (which means leaving Nature alone, and avoiding human-caused fire), and on the other hand, where there is danger of fire spreading towards settlement, the same managers are expected by society to minimise the risk of fire outbreak (in which control burning plays a major role). There is no easy solution to the dilemma, and although 'green' philosophies which oppose control burning have gained considerable community acceptance in recent years, after a major fire season, such as 1983 in Victoria and South Australia, or 1994 in New South Wales, the pendulum will swing toward a philosophy of fuel reduction.

There is also much that can be done to modify the loss potential from bushfire at the householder level. Activities which the individual householder may employ, and which can be crucial in areas exposed to bushfire danger (Fleeton, 1980), include:

- fuel reduction in the yard of a residence — collecting and composting dead plant material, removal of old timbers, etc.,
- appropriate storage of flammable materials such as fuels and firewood — *not* stacked next to, or underneath the house, or near a wooden fence which may spread fire,

- removal of litter such as dead leaves from gutters and other litter traps around the house,
- flush-screening of windows to eliminate airborne fragments of burning material from lodging in the window surrounds or entering the house,
- ensuring that sufficient garden hose is on hand, and there is adequate ladder capacity, to reach any point on the property.

Fire breaks *per se* are of limited usefulness in eucalypt forest environments, where spotting of wildfire is common. However, good fire management is based around the use of natural fire breaks such as topographic features devoid of, or poor in combustible material (rocky terrain, water bodies, etc.) or cultural features similarly devoid of fire-prone materials, such as roads. Fire breaks are most useful around high-risk areas such as saw mills, railway lines or main highways. In the absence of spotting behaviour, a fire break of approximately four times the maximum horizontal flame reach would suffice. Fire breaks also have a secondary function as refuges for humans and stock. They are however expensive to construct and maintain, even where herbicides are used.

Fire detection is principally by visual means. Fire detection towers are frequently located at high points in production forests and other strategic places. If a fire may be observed from two or more points, its position may readily be determined by triangulation. However, with large areas of land to manage, fragmented forest estate, and small staffs, bushfire management authorities in Australia are very reliant on the goodwill of neighbours and the public to report fires. People such as aircraft pilots who regularly observe the country also assist. Satellite remote sensing may provide information on fire in remote areas (Kaufman et al., 1990; Malingreau, 1990; Langaas, 1992).

Public education programs, and legal bans on outdoor fires on days of high fire danger, are used in an attempt to minimise the possibility of unintentional ignition, but fire managers are continually involved in forward planning, training of personnel, weather and fuel monitoring, and stockpiling of essential supplies for *fire suppression*, in which every effort is made to extinguish a fire in the minimum time and with minimum area burned. Fire suppression is the critical and timely component of good fire management, and uses trained firefighters (many of them being volunteers in the Australian context), specialised equipment, and para-military organisation. Aircraft may be employed, to 'bomb' the fire with water laced with fire-retardant chemicals, or to monitor the fire front — Matthews and Jessell (1993) report on an airborne infra-red scanner-plus-geographic information system capable of mapping a fire front (even through a heavy smoke cloud) and transmitting maps via mobile fax to fire chiefs in the field — and also earthmoving equipment to construct tactical firebreaks, especially in conjunction with back-burning operations. Time is of the essence since a fire may be prevented from developing its full potential with a timely suppression manoeuvre. Governments spend large sums on fire suppression activities.

In deciding on the most appropriate strategy for suppression of a fire in difficult terrain, rapid deployment of crews at the point where the fire may be weakest or most easily brought under control is vital. Resource management in fire-prone regions therefore includes the detailed assessment of fire potential for alternative weather and landscape scenarios, which requires the ability to predict the consequences of site-specific management actions on fire characteristics for various topographic sites, vegetation types and arrangements. Expert systems have been developed for some areas (see, e.g., Davis et al., 1986), and another approach uses fire behaviour prediction models, such as FIREMAP or ASH FRIDAY, or Wallace's (1993) fire ellipse model under typical fire season meteorological conditions. FIREMAP is a simulation system designed to estimate wildfire characteristics in spatially non-uniform environments and simulate the growth of fire in discrete time steps. It integrates Rothermel's fire behaviour prediction model (Rothermel, 1972, 1983) with a raster-based geographic information system. The outputs can be displayed as digital maps (Vasconcelos and Guertin, 1992). ASH FRIDAY (Chapman, 1992a) is an educational simulation model which also used gridded data on terrain, vegetation types, and cultural features, readily compiled for any specific area, with randomly located fire start. Fuel load is computed for each vegetation type on the basis of time and weather since the last fire. Fire progression then depends on the above parameters, plus weather (drawn from a database of typical fire weather sequences), and management actions taken by the 'fire managers' (students). Complete recalculation of conditions is made for every minute of (simulated) time and the progress of the fire is displayed on a computer screen map, along with data on the fire and area burnt.

Conclusion

The forests and woodlands of the 'green crescent' of southeastern Australia, in particular, create hazardous conditions for wildfire, with the volatile oils, flammable bark, and abundant woody litter of the eucalypts, and, in many cases, dry, sclerophyllous shrub understoreys. And the interface between these fire-prone, fire-adapted environments and urban settlement continues to expand. The desire to live in an area of high en·/·ronmental quality, whether the seashore, the mountains or the forest, is one which we would surely only wish to encourage. There are good reasons for wishing to live at the margins of the bush, but the price is the risk of occasional catastrophic damage. We might ask whether that risk could be minimised, especially in new areas, by planning and building controls. We have in place many planning controls based on environmental amenity and site suitability, but do they adequately address the bushfire hazard? The imposition of building codes in the interests of structural integrity is accepted almost without question by homeowners: could not fire-resistant construction be specified in areas of known fire risk? And what of the insurance system? Safe drivers and dwellers in low-crime risk areas are 'rewarded' with lower premiums: incentives for fire-resistant construction, or conversely, a bushfire risk loading, may be appropriate. Local governments and fire management organisations involved provide high-quality information to householders on the degree of hazard and on means of reducing risk to individual properties, and carry out fuel reduction burning in their areas of responsibility. Fuel reduction strategies cannot prevent bushfire, however; their aim is to reduce probability of ignition, or, if ignition occurs, to produce a slower, lower intensity, more controllable fire than would otherwise be the case. But the attempt at reduction of fuel may be quite ineffective — forest which supports a conflagration in a time of high fire danger may refuse to ignite, or burn only in patches, under conditions in which fuel reduction burning is safe. The costs of complete fuel minimisation and fire exclusion in the affected areas would, over time, probably exceed the value of property being protected. With the best of preparedness (we cannot control the weather), it is inevitable that at some time a fire beyond our capacity to control will occur, especially given the fact that most are started by people. Residents of fire-prone environments must be prepared to make the effort of understanding those environments, and must accept the consequences of their locational decisions: the possibility of catastrophic loss in the event of an 'unstoppable' fire, and the necessity of both physical and financial measures to insure against loss.

The house was shelter. It had been designed to withstand bushfire, and even had its own gravity-fed tank to supply water during a power failure. If they survived the thermal pulse, they could probably save the house. With heat and noise, however, came doubts. A tidal wave of flame twenty meters high crashed upon them. The curtains started to smoulder, windows cracked, the wooden window frames in the study burned through. A bowl of pancake batter on a kitchen table cooked. The small band furiously swatted out tiny fires that ignited and passed buckets of water from a bathtub. The heat was a hundred times greater than the brightest sunlight. Yet they lived. The house endured. And Ash Wednesday experienced one of those epiphanies that transform a story of survival into a moral universe, a world of conscious choice, of acts that express a relationship of humans to the environment around them. (Pyne, 1991: 416–17)

3

Storms and their outcomes — The restless atmosphere

Viewed from space, the earth is surrounded by a soft blue envelope, the intricate motions of which are picked out by lacework in cloud. On the surface, some of these same motions unleash forces beyond human capacity to cope, in the form of storms. Any disturbance of the atmosphere, accompanied by wind, rain, snow, hail, or lightning and thunder, is termed a storm, but it is the storms that produce extremes of these phenomena that are characterised as hazards. Hence, extreme wind, rain or other forms of hydrometeors, or lightning, may arise from a variety of different atmospheric disturbances, but the principal hazardous storm types are tropical and extra-tropical cyclones, severe convectional thunderstorms, and tornadoes. Tropical cyclones are estimated to cause average losses of $260 million per year in Australia, and other storms to cause losses of $200 million (AWRC, 1992). In the discussion which follows, atmospheric hazard phenomena are organized into separate categories for the sake of clarity, although many overlap — wind, for example, accompanies most storms, and lightning is sometimes generated by volcanic eruptions in addition to normal thunderstorms. Because windstorm accompanies many atmospheric disturbances, it is discussed first, before the principal storm types, with their commonly associated secondary hazards: storm waves and storm surge with tropical cyclones; lightning and hail with thunderstorms. Temperature extremes, sometimes, though not always associated with storms, deserve a separate section at the end of the chapter.

WIND

The hazard of windstorm may result from tropical and extra-tropical cyclones, tornadoes, thunderstorms, monsoons, katabatic winds, and firestorms. It is the most pervasive of all natural hazards as there is no location on earth where wind does not blow, and according to Munich RE (1990) accounted for about 50 per cent of the major natural disasters recorded between 1960 and 1989. Storm waves (discussed below), a coastal and sea hazard, are a secondary product of strong winds.

The impact of the wind hazard is principally on buildings, and the effect of wind on a building may be understood by considering the structure as consisting of five basic surfaces (four walls and a roof). The overall pattern of air flow produces an inward-acting pressure across the windward wall and outward-acting pressures across the two side walls, the leeward wall and the roof (figure 3.1a). If the building is enclosed and contains air at normal atmospheric pressure, there will be outward-acting pressures across four of the five surfaces as a result of reduction of outside pressure caused by wake turbulence and by acceleration of the air flow as it travels the longer distance around the structure (a manifestation of the Bernoulli principle, which predicts that fluid pressure will be reduced wherever the velocity of flow is increased — the same principle is used in the design of aeroplane wings to produce lift). Locally, the air flow is also interrupted by sharp edges at eaves, roof ridges and roof and wall corners. The air flow separates from these edges, creating relatively low pressures immediately downstream, which may result in locally severe outward-acting forces. In addition to flow separation and Bernoulli phenomena, additional pressure differences across roofs and walls occur when openings appear in the structure, either by design or as a result of component failure (figure 3.1b). The aerodynamically poor configuration of most conventional buildings invites relatively large outward-acting pressures near roof ridges, roof corners, eaves and wall corners, and the openings in buildings can compound outward-acting pressures. Wind, acting alone, can thus produce outward pressures across all surfaces except the windward surface: the wind-induced pressures can produce the appearance of building explosions during severe wind storms.

Pressures due to wind are not steady, but fluctuate in time, because the wind itself is turbulent, with random variations in wind speed, known as gusts. The greater the surface friction of the landscape, the stronger the turbulence at sea, gusts rarely exceed 1.5 times mean wind velocity, while over rough terrain they may reach 3 times mean wind velocity (Munich RE, 1990). Moreover, the structure itself may be capable of inducing fluc-

Figure 3.1a Overall wind effects on building surfaces.

Figure 3.1b Wind effects on building surfaces due to openings.
Source: Minor and Mehta (1979). Reproduced by permission of American Society of Civil Engineers.

.tuating wind forces, either from movement of the structure, or from the formation and discharge of vortices from its surfaces. The oscillatory response to unsteady wind forces may be of sufficient amplitude to cause overstress or collapse. Even where amplitude is limited, it may be sufficient to cause eventual collapse from fatigue (Scruton, 1981).

Contrary to popular opinion, masonry walls are not proof against wind damage. It is entirely possible for wind-induced pressure to push a masonry wall inward or outward, depending on wind direction and the character of the wall openings. Where a roof system is supported by the walls, the collapse of a single load-bearing masonry wall may cause catastrophic failure of the entire building.

The lightweight cladding of many low-pitched domestic roofs is especially prone to wind damage. Water damage from rainfall entry subsequent to wind damage to buildings is a major cost factor, especially when the winds are generated by tropical cyclones. Windborne missiles can cause significant damage in wind storms, and in extreme wind conditions, failure of one building or its components may create hazards downwind. Building components can become windborne missiles, perforating and damaging otherwise adequately engineered buildings.

Assessments of wind induced damage to structures reveal that:

- conventional buildings, especially housing, are often damaged by winds as low as 33 m s^{-1} (~120 km/h), and most of the damage to conventional buildings produced by windstorm is caused by winds of less than 55 m s^{-1} (~200 km/h). Extremely small areas are affected by winds exceeding 75 m s^{-1} (270 km/h);
- small increases in degree of engineering attention can produce very large dividends in increased wind resistance;
- structure geometry and orientation are important determinants of wind resistance;
- certain types of structures (e.g., mobile homes, frame housing, unreinforced masonry buildings) are very weak where wind resistance is concerned.

Clearly, the most effective management strategy for the wind hazard is appropriate building design which takes account of maximum expected wind gusts. In technologically advanced countries, these data are usually codified into statutory building codes.

Wind shear and microbursts

Low altitude wind variability, or wind shear — a dramatic change in wind speed, or direction, or both, between two adjacent air currents — has long been recognised as a hazard to the landing and take-off of aircraft, especially when it occurs in the lowest hundred metres or so of the atmosphere. Aircraft must penetrate this zone on take-off or landing, and sudden loss of lift and/or altitude can be critical. Wind shear may be either vertical or horizontal: strong wind variations over horizontal distances as great as several kilometres can cause particular difficulties for commercial aircraft. Wind shear may be encountered under a variety of weather situations:

- along fronts. Winds always change across a frontal system, and there can be marked discontinuity in speed and direction leading to wind shear;
- at low-level temperature inversions which form during clear calm nights. Winds within the cold surface air layer usually are calm or very light, but winds in the warmer air above may be stronger, leading to wind shear and turbulence at the inversion layer;
- in association with cold-air downdraughts from thunderstorms. Pilots are trained to avoid thunderstorms because of associated severe wind variability.

Cold-air downdraughts from thunderstorm cells, known as *microbursts*, are intense and short-lived — normally not longer than about five minutes. The air carried downward issues from the leading edge of the (sometimes immature) thundercloud, impacting the ground with velocity up to 100 km/h and then bursting radially outward along the ground in a shallow layer often only about 100 m thick, and usually less than 4 km in diameter. Violent turbulence may develop between the warm air rushing into the convectional updraught of the thunderstorm cell and the downrushing colder air.

If a landing aircraft on its glide path encounters a microburst, it will first experience a strong headwind, which will increase its airspeed but decrease its groundspeed (forward motion relative to the ground). The initial result is unwanted lift, which will induce the pilot to put down the nose of the aircraft. However, when the aircraft leaves the headwind zone of the downdraught, there is abrupt loss of lift because the wind direction changes abruptly and becomes a tailwind, sharply reducing airspeed: lift may then be inadequate, and the aircraft may stall and crash. Microbursts have been implicated in crashes of quite large commercial aircraft.

TROPICAL CYCLONES

A tropical cyclone (figure 3.2) is defined by the Australian Bureau of Meteorology as a non-frontal, synoptic scale, cyclonic rotational (i.e., clockwise in the southern hemisphere), low pressure atmospheric system of tropical origin, in which ten-minute averaged mean winds of at least gale force (63 km/h, or 34 knots) occur, with the belt of maximum wind being in the vicinity of the centre of the system. A tropical depression, on the other hand, is defined as a closed, rotational storm system of tropical origin in which mean winds at any point do not exceed gale force. Tropical cyclone winds may reach sustained speeds of more than 240 kilometres per hour (130 knots). Tropical cyclones usually develop from a tropical depressions which form at sea, although they have been known to form from tropical depressions which originated inland and moved out to sea. The term 'tropical cyclone' is also used in the Indian region, but in the northwest Pacific they are referred to as typhoons, and in the Americas as hurricanes. Specifically, in North American usage, a severe tropical storm is called a hurricane when the maximum sustained windspeeds reach 120 kilometres per hour (65 knots).

Tropical cyclones, unlike less severe tropical storms, are generally well organised and have a circular wind pattern with winds revolving around a centre or eye (not necessarily the geometric centre). The eye is an area of low atmospheric pressure, light winds and, sometimes, relatively clear skies. Atmospheric pressure and windspeed increase rapidly with distance outward from the eye to a zone of maximum windspeed, which may be anywhere from a few to a hundred kilometres from the centre. The rotational character of the storm means that an observer on the ground, in the path of the storm, would experience wind of increasing velocity as the eye approached, but as the eye moved overhead, the wind would drop to almost a calm. After the eye passed, wind would rapidly pick up again, in the opposite direction to the earlier winds, as the second wall cloud

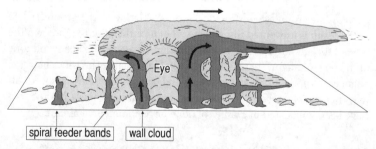

Figure 3.2   Cross section through a typical tropical cyclone. In this schema, winds are moving towards the observer, out of the page, to the right of the eye, and away from the observer to the left of the eye. In addition to the rotary movement, however, there is also vertical air movement, especially in the eye wall cloud and the spiral feeder bands, resulting in high level outflow in the cirrus region, at about 12 000 m in our cross-section.

moves over (see figure 3.3). From the zone of maximum wind to the periphery of the tropical cyclone, the windspeed continues to decrease, but the pressure increases. Central pressure (atmospheric pressure within the eye) invariably rises, and the strongest windspeeds decline, when a tropical cyclone passes from sea to land. The overall diameter of tropical cyclones varies greatly, from less than 300 km up to 3000 km. Generally, for tropical cyclones of a given size, the lower the central pressure, the higher the windspeeds. An index of the size of the storm is given by the radius of maximum winds and the speed of forward motion of the storm system. The central pressure is the best single index for estimating the storm surge potential of a tropical cyclone.

Tropical cyclones are usually spawned in the warm waters of the Pacific, Indian and North Atlantic oceans between about 20°N and 15°S, although rarely in the zone between 5°N and 5°S where the Coriolis effect necessary for rotational storms is weakest. Tropical cyclones are not known to cross the equator, and are not known from the coasts of sub-equatorial South America. Frequency is highest in late summer–autumn because tropical cyclones depend on warm sea surface temperatures (with a normal lower limiting value of 26°C), which lag behind the sum-

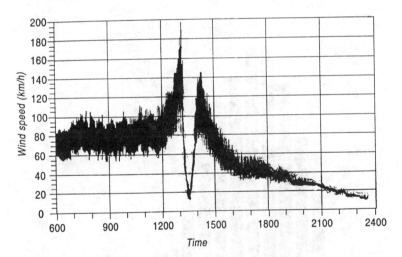

Figure 3.3 Anemometer (wind speed recorder) trace of a typical tropical cyclone passing more-or-less directly over the recording instrument. Strong winds may be experienced for many hours before the tropical cyclone makes landfall. Wind speeds increase gradually as the storm nears the instrument, and then rapidly as the eye approaches. Maximum wind speeds are recorded in the eye wall, but inside the eye, conditions are almost calm. Following the eye, wind speeds increase rapidly again, but now in the opposite direction. The wind speed decreases again as the tropical cyclone moves away.

mer solstice by 2–3 months. Worldwide, there are about 80–100 tropical cyclones per year, and in the Australian tropical cyclone region (105°E to 165°E and ≥ 5°S), there are on average about eight (figure 3.4).

A tropical cyclone gives rise to a complex of impacts (figure 3.5). The immediate and long- term impact of tropical cyclones are related to their magnitude and frequency, the economies and mode of life of the affected countries, and the understanding of the tropical cyclone phenomenon by the affected communities and their willingness and capacity to act on that information. The regions of the world most vulnerable to socioeconomic impact from tropical cyclones are the tropical and subtropical areas of: fertile, densely populated deltas, especially those in zones of high tidal range (e.g., Ganges-Brahmaputra); small islands, especially those heavily dependent on agricultural economies (e.g., Philippines, Pacific and Caribbean island nations); and, well-populated coastal areas with high levels of investment, especially if low-lying (e.g., parts of the coast of Queensland and of the southern states of USA, Hong Kong).

In Australia, a string of seven tropical cyclones in 1967 caused damage estimated at $300 million to the Gold Coast of Queensland: much of the damage was due to wave erosion of the coastal dunes and structures built

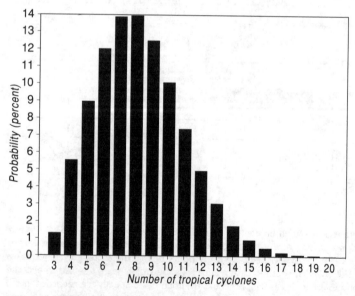

Figure 3.4 Probability of tropical cyclones in the Australian region.
Based on Poisson model fitted to updated and de-trended data of Lourensz (1981).

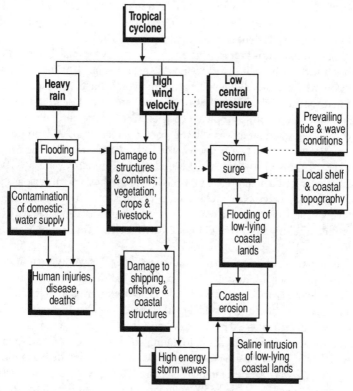

© D. M. Chapman

Figure 3.5 The impacts of tropical cyclones result directly from high winds, storm surge, heavy rain and flooding, and secondarily from the high energy storm waves generated by the winds from the tropical cyclone.

on them (Delft, 1970). Heavy rainfall from tropical cyclone Wanda in January 1974 resulted in one of the worst flood disasters in Australia, when the rainfall caused near-record rises in the Brisbane River and its tributary the Bremer. Twelve people lost their lives, and damages were estimated at $178 million (Chamberlain et al., 1981b). However, the most damaging tropical cyclone to date has certainly been tropical cyclone Tracy, which brought the northern city of Darwin to the world's attention by largely demolishing it on Christmas Day 1974 (see Bureau of Meteorology, 1977, Chamberlain et al., 1981a, and several papers in Heathcote and Thom, 1979). Tropical cyclone Tracy was to test the newly formed Natural Disasters Organisation to its utmost (Stretton, 1976), and became a social reference point in Australia, especially with regard to natural hazards (for example, a map of tropical cyclone risk reproduced by Oliver (1986) depicts 'recurrence intervals' in terms of 'wind damage equivalent to tropical cyclone Tracy').

By global standards, tropical cyclone Tracy was a very small but intense storm, the diameter of gale force winds being only about 100 km (in comparison to some north Pacific typhoons which have corresponding diameters of around 1500 km), with a central pressure close to the mean value observed in other tropical cyclones. Central pressure of tropical cyclone Tracy at landfall was 950 millibars (hectopascals), and although maximum wind velocity was not recorded (the anemometer having been broken by the storm), Bureau of Meteorology estimates were for 10-minute averaged maximum mean wind of 140–150 km/h, with peak gusts of 217–240 km/h. Tropical cyclone Tracy occurred at a time of neap tides at Darwin, and although the maximum storm surge was associated with a high tide, the surge of 1.6 m did not create problems (Bureau of Meteorology, 1977).

There have been many tropical cyclones of greater severity than tropical cyclone Tracy in Australia, and had it crossed the coast 50 km or so either side of Darwin, its impact would have been unremarkable. However, the eye of the storm passed directly over Darwin, and left no part of the city undamaged — 90 per cent of housing was destroyed or rendered uninhabitable; 65 people died (including 16 lost at sea), creating an unusual peak in the Australian record of deaths from storm disasters (figure 3.6); 140 were seriously injured, about 1000 received minor injuries; and total damages exceeded $500 million (Leicester and Reardon, 1979; Stretton, 1979). Failure rate of properly engineered buildings was small (about 3%) compared with that of non-engineered structures (about 60% — Stark and Walker, 1979): simple anchorage systems for roofs, and the use of 'cyclone bolts', which anchor the frame of a building to its foundations — measures which add a mere few hundreds of dollars to the cost of a home if installed during construction — add considerably to the wind resistance of frame buildings (Leicester and Reardon, 1979).

In the Americas, Hurricane Hugo, of September 1989, and Hurricane Andrew, of August 1992 will long be remembered for damages of about US$10 billion and US$30 billion respectively, the latter the most costly natural disaster in the history of the USA to the time of writing. Hurricane Hugo moved right across the state of South Carolina, and exited with winds still at or near hurricane strength. It resulted in the greatest forest disaster in the history of the state: 1.8 million hectares of timberland were damaged, with damage in six counties being close to 100 per cent, and water-damaged forest habitats will require decades to recover

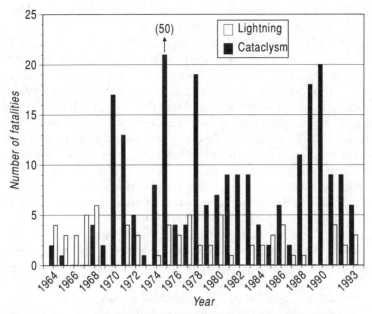

Figure 3.6 Thirty years of fatalities in Australia from Cataclysm (Australian Bureau of Statistics category for injury from tropical cyclone, tornado, flood, hail, landslide, etc.), and from lightning. Forty-nine of the fifty deaths reported in 1975 were from tropical cyclone Tracy on 25 December 1974, but the figure does not include 16 people lost at sea, presumed drowned.
Data from Australian Bureau of Statistics.

(Sheffield and Thompson, 1992). The impact of the hurricane revealed a wide spectrum of building and management practices ranging from 'excellent to execrable' (Miller, 1990), and JCR (1991) reported that Hurricane Hugo caused building damage far in excess of what 'should have' occurred: much of the damage inflicted on the built environment (15 000 homes destroyed) was attributable to failure to implement mitigation measures recommended by engineering professionals more than twenty-five years earlier. In the aftermath, Miller (1990) observed that neither lending institutions nor insurance companies had taken pains to exert their considerable leverage over the real estate market to ensure that mitigative measures were implemented to reduce wind damage. Shore protection structures for the most part were not designed to withstand a storms of the strength of Hurricane Hugo, and in any event, most were overtopped by the storm surges. Similar outcomes (and criticisms of building quality) were experienced in southern Florida and Louisiana when Hurricane Andrew moved through in 1992: more than 80 000 buildings effectively demolished, another 50 000 severely damaged but able to be used, and 160 000 people homeless.

Severe damage has also been caused by extra-tropical cyclones. For example, in 1987 a cyclonic storm crossed over southern England and felled almost twice the volume of broad-leaved trees that are harvested commercially in any typical year (2 million cubic metres). The volume of conifers downed (about the same amount) was about half the average annual commercial harvest of such species. Some 15 million trees were blown down. Tree loss assumes great significance in the UK, which is one of the least forested countries in western Europe. Wind gusts exceeding 90 knots (165 km/h) were reported — at some locations, the return period for wind gusts of that magnitude is in excess of 500 years. Earlier, in 1953, a North Sea depression with central pressure as low as 968 mb caused winds of 100–130 knots (185–240 km/h), gusting to at least 150 knots (280 km/h), with storm surge at the Netherlands coast of 3 m, and 2.5 m in the Fens area of SE England (Lamb, 1991). The arrival of the surge coincided with a high spring tide: over 2000 people drowned in the Netherlands, and there was catastrophic damage to stock and property, with 25 000 square kilometres of land flooded by sea water.

Management of tropical cyclone impact is discussed in the specific sections dealing with wind, storm surge, etc., but it must be noted here that, unlike some other natural hazards, establishment of an effective warning system for tropical cyclones is a realistic goal, and most areas subject to them are covered by warning systems. A number of tropical cyclone warning centres cover the Australian region, each of which maintains a continuous watch throughout the cyclone season by analysing satellite and radar signals, surface observations (from ships at sea, automatic weather stations offshore, and from weather stations on land), and upper atmosphere wind and temperature fields (balloon, radio-sonde, high flying aircraft, satellite). Cyclone genesis and movement are monitored and a variety of warning types may be issued (Wilkie, 1984):

1 Information — possible Tropical Cyclone within 48 hours; normally part of regular news broadcast, and at same time,

2 Tropical Cyclone Watch — advice of possible TC issued to emergency services;

3 Tropical Cyclone Warning — declared as soon as gales (>63 km/h) expected within 24 hours; and

4 Storm Tide advice if storm surge expected within 24 hours.

Warnings of type 3 or 4 are updated every three hours if conditions warrant, and every hour if the centre of the TC is close. Content of warning messages includes prediction of position of cyclone, wind strength, state of tide, and likely areas of heavy rain and flooding. An informative brochure freely available in cyclone-prone areas (*Know your cyclone warning system*, issued by Bureau of Meteorology), gives basic information about tropical cyclones, about the types of warning messages to expect, and advice on suitable responses.

However, the high mobility of most coastal populations in developed countries, the increasing concentration of populations in coastal areas, and the attractiveness of the coast as a second-home or retirement location means that many people exposed to the tropical cyclone hazard have had no experience with the phenomenon, and find it difficult believe that they may be subjected to one. Many people do not take warnings seriously. In addition, having in many cases moved to the coast because of its climatic advantages (who looks at real estate in foul weather?), they find it difficult to comprehend that the climate could 'turn against them'. Surveys of residents at the Gold Coast, Queensland, a highly developed area at risk from tropical cyclones, by the author and by Hobbs and Lawson (1982) have revealed a low level of both hazard awareness and preparedness, and a strong tendency to explain away any perceived risk ('They only happen every fifty years', 'The Council always takes care of it').

Storm surge

The term 'storm surge' is used to describe the elevation of ocean water level due to the action of cyclonic storms. Abnormal rises in water level in nearshore regions will not only flood low-lying terrain, but provide a base on which high waves can attack the upper part of the backshore and penetrate farther inland. Flooding of this type combined with the action of waves can cause severe damage to low-lying land and backshore developments. Coastal erosion and loss of structures built on coastal sands, displacement of stone armour units of groynes and breakwaters; cutting of new inlets through spits and barrier beaches, and shoaling of navigational channels can often be attributed to storm surge and the surface waves superimposed upon the raised water level. Moreover, surge can increase hazards to navigation, impede vessel traffic, and hamper harbour operations. Apart from normal astronomical tides, which are predictable, the factors that may be responsible for elevated water level during the passage of a cyclonic storm are:

- atmospheric pressure differences (the 'inverse barometer' effect). The rise in water level due to the reduction in atmospheric pressure from normal is almost exactly 1 cm for every millibar (hectopascal) drop in pressure.
- onshore winds — winds alone may be responsible for significant changes in water level during a storm. A wind blowing over a body of water exerts a horizontal force on the water surface and induces a sur-

face current in the general direction of the wind, which may cause water to 'pile up' at the shore.

- wave setup from storm waves. Wave setup occurs between the wave break-point and the beach, and is caused by the water from broken waves being ponded against the shore by incoming waves. It can be as much as 10–20 per cent of incident wave height.
- rotation of the earth (Coriolis effect) — the Coriolis effect deflects a flow to the left in the southern hemisphere, and to the right in the northern hemisphere. It is weakest near the equator, but on the coast of eastern Australia for example, a strong southerly or southeasterly wind blowing more-or-less parallel to the coast would generate a current which would 'pile up' against the coast, due to leftward deflection by the Coriolis effect.
- rainfall — heavy rainfall is frequently associated with tropical cyclones and may result in a local rise in water level in estuaries and lagoons due to freshwater runoff. In addition, there is the ponding effect of a lens of freshwater runoff superimposed on the denser salt water of the ocean.

Nowhere in the world is there any other location as prone to the effect of storm surge as is the coastal zone of Bangladesh. Murty and Neralla (1992) show how the particular topography of the Bay of Bengal and its shallowness, together with the large tidal range, make the storm surges on that coast more dangerous than those anywhere else on earth. The combination of tropical cyclones with the normally large astronomical tide, funnelling coastal configuration of the upper Bay of Bengal, the low, flat terrain broken by river distributaries, and extremely high population density set the scene for major storm surge disasters in 1737, 1864, and 1876, in each of which over 100 000 died by drowning or in the subsequent cholera epidemics, and in 1991, when 138 000 people were drowned (Katsura, 1992) and perhaps 10 million made homeless. However the 1970 event ranks as perhaps the deadliest tropical cyclone in history (Frank and Husain, 1971), killing over 300 000 people (some observers put the total at 500 000) and directly affecting 4 million more, plus countless animals. Reports of the maximum surge height differ, but the (probably conservative) Water and Power Authority recorded a maximum depth of inundation of 16 feet (4.9 m) at two surge gauges on Hatia Island in the Ganges delta. Sixty-five per cent of the total fishing capacity of the country was destroyed by the storm (and 80% of the meagre annual protein intake of residents of Bangladesh comes from fish!). Poignantly, one of the outcomes of the storm was the cleaving of East Pakistan from Pakistan to create the new nation of Bangladesh — a nation born in travail.

In Australia, the sparsely populated nature of the tropical coasts explains the low incidence of disaster due to storm surge — the most serious being in 1899, when a pearling fleet in Bathurst Bay (N. Qld) was destroyed, with 300 dead, by a surge reported at 12 m (Whittingham 1963). Other surges of up to 7 metres have been recorded at remote parts of the coast, and at both Townsville and Mackay in Queensland, recorded surges of almost 4 m (Hopley, 1974) present a warning against settlement in low-lying parts of the coast where storm surge may be significant.

Katsura (1992), Murty and El-Sabh (1992), and Khalil (1992) present a number of potential mitigation strategies for storm surge, some quite fanciful, and conclude that there is most merit in surge barriers (dykes) and elevated surge shelters, while emphasising the importance of dense coastal forest as a friction-generating device which hinders surge penetration. Khalil draws particular attention to the $^1/_2$ million-hectare tract of mangroves, known as the Sunderban, in the southwest of Bangladesh, which has proven to be an efficient form of protection against storm surge. As there is usually a warning of several hours before arrival of a storm surge, people may sometimes be evacuated from affected areas. In the USA, the National Weather Service has commissioned special Storm Evacuation Maps for parts of the south and Gulf coasts, showing areas likely to be flooded by different surge heights, and principal evacuation routes. However, the time needed to achieve successful evacuation sometimes exceeds the maximum possible advance warning time. As traditional 'horizontal' evacuation methods become less feasible, 'vertical' evacuation into properly designed high-rise buildings capable of resisting hurricane stresses is becoming an appealing alternative.

Storm waves and coastal erosion

The risks associated with living along a sandy seashore are comparable in some ways to those of living in a riverine floodplain, near an earthquake fault, or close to a volcano; all these areas are subject to extreme natural changes and carry the possibility of eventual catastrophe for residents (Williams et al., 1991). All beaches exhibit dynamic responses to variations in the natural suite of processes: waves, tides, currents, surges, and rainfall all induce mobility in beach sediments. The hazard of coastal erosion is historically recent, being associated with the upsurge of coastal population density in many parts of the world in the twentieth century — if sandy shorelines were neither settled nor otherwise used by us, coastal erosion would be principally a matter of scientific curiosity. The population of Australia is highly concentrated at or near the shore, especially in the environs of the capital cities, all of which are on the coast: over a quarter of the nation's population lives within 3 km of the shore, and three-quarters within 40 km (Chapman, 1980). Locations close to the beach, especially the crests of frontal dunes, are highly favoured. However, it is those very same locations that have proven vulnerable when major storm wave erosion bites into the shoreline (figure 3.7).

The sandy beach is actually a prism of sand, as it has depth as well as length and breadth, and the volume of sand in the prism is the balance of a budget of sediment supply and loss:

Sources	balanced by	Sinks
Fluvial supply		Loss to inlet or estuarine sinks
Supply from inner shelf sands		Offshore loss
Biogenic supply		Solution and abrasion loss
Supply from cliff or dune erosion		Loss to dune migration inland
Supply from beach nourishment		Loss to beach mining
Longshore transport supply		Longshore transport loss

Sand may be gradually lost from the prism over a long period, due to imbalance in the supply/loss relationships outlined above, but the deficiency is usually made apparent only when the beach is attacked by storm waves. The energy in ocean waves is proportional to the square of the wave height and, compared with the energy delivered to the shoreline by normal swell waves, the energy of storm waves can be immense: at the Gold Coast (Qld), for example, normal wave power averages 9.59 kW m^{-1} (i.e., 9.59 kilowatts per metre length of wave crest), but during major storm events it may exceed 300 kW m^{-1} (Chapman, 1981). During a major storm, sand is eroded from the upper, normally dry, beach and from the frontal dune, and transported into the surf zone, where it is deposited in a 'storm bar' at or near the break-point of storm waves. The amount of material taken offshore from the upper beach may amount to several tens or even hundreds of cubic metres of sand per running metre of beach, and result in substantial cut or even destruction of the frontal dune. Storm events leading to disastrous erosion occur on average once in several decades, and usually involve a cluster of storms within a short time (Chapman et al., 1982).

Management of the coastal erosion hazard has been attempted by *avoiding the hazard* by using land-use control to create buffer zones (called setbacks in the USA) for development; *modifying the hazard* by construction of revetments to prevent storm waves eroding the backshore,

Figure 3.7a and 3.7b Crests of frontal dunes have been favoured places for construction of seaside homes, as a location in which to enjoy all the advantages of the seashore. However, over the long term, the frontal dune becomes, from time to time, part of the active zone of the beach.
Photograph (a) © *The Boston Globe*. Used by permission.

building breakwaters to prevent storm waves reaching the shore, installing groynes to trap sand and augment the swept prism on the updrift side, or beach nourishment/dune building to augment the sediment budget of the beach so that its role as a 'shock absorber' of storm waves is enhanced; and *preventing loss* by strengthening buildings and/or ensuring that they are on strong, deep footings, by relocation of buildings from the hazard zone, or by emergency evacuation of people or portable property from the hazard zone. Losses have also been shared through insurance or government handouts, or borne by individual property owners. Thieler et al. (1989) emphasise the importance of good foundations and minimal modification of the shore zone in connection with protection from storm wave and surge damage. Their study showed damages to be consistently lower where coastal forest, natural dunes, and wide beaches (all of these sometimes preserved as a result of setback regulations) protected housing, but that seawalls, buildings, and rubble barriers caused beach narrowing, exotic vegetation provided less protection than native forest, and modifications to the dunes provided foci for attack: notches in dunes, and roads at right angles to the shore, became overwash passes for waves and storm surge. Most of the seawalls in the area studied by Thieler and co-workers were small ones, and were simply overridden by storm surge, with wave attack consequently focused on the first row of development.

THUNDERSTORMS

Thunderstorms are common in the tropics, less so in the mid-latitudes, and rare beyond the Antarctic and Arctic circles. In the mid-latitudes, most thunderstorms occur in the summer–autumn half year, and, like tropical cyclones, also accompany warm ocean currents. A thunderstorm is a deep, convective cumulonimbus cloud (or agglomeration thereof) which produces lightning (and, of course, thunder), heavy rain and/or hail, strong surface outflow of cool air, and, on very rare occasions, a tornado. The typical thunderstorm cell is deep (from several hundred metres to perhaps 12 000 or 15 000 metres above the ground), cylindrical or slightly hour-glass shaped, and capped by an anvil-shaped top caused by air diverging from the terminal updraughts of the storm (see figure 3.8). Thunderstorms are usually initiated by buoyant lifting of air, which results from air above a particularly warm land (or water) surface becoming heated until it is warmer than its surroundings. Thunderstorms may also develop from uplift over mountains, or lifting of air along an air mass discontinuity — a squall line, as in a cold front — the phenomenon known as dynamic lifting.

A thunderstorm is a 'thermodynamic machine', driven by the latent heat of condensation or sublimation (heat released when water vapour is converted into cloud droplets or ice crystals). Successive 'thermals' of heated air develop into cumulus clouds that grow higher and higher until a fully fledged thunderstorm forms. As the air rises, raindrops increase in both number and size, particularly when the freezing level in the atmosphere is reached by the vertical build-up of the cell, because condensation of water vapour directly to ice crystals (sublimation) can occur faster, and at lower atmospheric humidity, than is required for growth of cloud droplets. Precipitation does not occur immediately, however, as the updraughts initially support the raindrops or ice crystals. Updraughts in the mature storm may reach 25 m s^{-1} (~90 km/h) or even higher. When precipitation does begin, it is accompanied by strong downdraughts (which eventually supplant the updraughts), with velocity

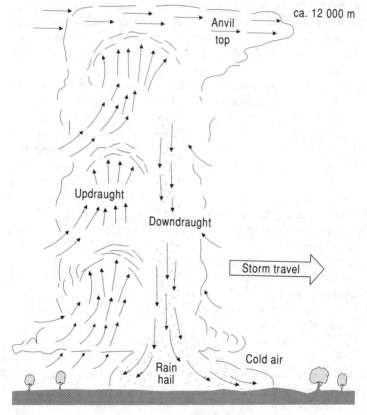

Figure 3.8 Structure of a typical thunderstorm cell.

of 15 m s^{-1} (~55 km/h) or more, diverging out from the precipitation area.

The life cycle of a typical thunderstorm is only 1–2 hours, and many thunderstorms mature, 'rain out' and die soon after reaching maximum vertical extent. The mature stage is characterised by precipitation reaching the ground, and by updraughts and downdraughts co-existing side by side, as shown in figure 3.8. Severe *supercell* thunderstorms — the type which sometimes give rise to tornadoes — develop a cell or two of extraordinary magnitude in the mature stage, typically with updraughts that penetrate well into the stratosphere (perhaps to 18 000 m or more), in an environment marked by extreme wind shear. The latter characteristic allows the precipitation downdraught to form adjacent to (rather than within) the updraught, and thus promote a quasi-steady state mature stage which can persist for some hours. Severe storms are also characterised by a weak cyclonic rotation of the air column, which may be associated with the generation of a tornado. In *multicell* thunderstorms, moist low altitude environmental air may be lifted by outflowing cold air from the storm, and the storm thus provides its own lifting mechanism. Although individual thunderstorm cells move with the mean wind, the multicell ensemble is not so restricted, as the storm may propagate stepwise in the direction of new cells as they are formed. Such a storm gave rise to the Sydney thunderstorm disaster of 21 January 1991.

On that Monday, atmospheric conditions were very unstable over eastern New South Wales — the result of warm, moist air at the surface and a 'cold pool' of air at middle levels of the atmosphere. Thunderstorms had started to form on the highlands over the eastern part of the state by midday. During the afternoon, a large multicell thunderstorm which had formed on the ranges about 50 kilometres southwest of Camden moved northeastwards through the Sydney metropolitan area at about 50 km/h. The storm lasted for about three hours, finally decaying off the coast south of Gosford.

As the storm moved through Sydney suburbs, it caused localised areas of intense damage, particularly in the upper North Shore–Duffy's Forest areas, due to wind gusts estimated at up to 230 km/h, hail up to 7 cm in diameter, and extremely heavy rainfall. The rainfall at a number of locations exceeded the 100-year Recurrence Interval rainfall for periods of 20 minutes. As the area of most severe impact was a part of Sydney noted for its fine trees, there were many snapped or uprooted trees, and broken branches, which in turn damaged houses, motor vehicles, and so on. There was also direct wind and hail damage to buildings, with many houses unroofed and some even blown off their foundations. Three large electricity transmission towers were also felled by wind. Water entry to buildings damaged by wind and hail was responsible for a significant proportion of the damages bill of $300 000 000. Hundreds of residents were displaced from their homes, and thousands more reverted to nineteenth-century pioneer lifestyles in the absence of electricity, telephone, and, in some cases, adequate shelter, for weeks afterwards. The clean-up and reconstruction process was noteworthy for the good humour and *esprit de corps* displayed by residents, as well as for the mountains of dead tree material collected by the affected local governments. Trees were completely defoliated by the hail, and this, with bark and structural damage, resulted in the subsequent death of many — as this is being written, three years after the storm, the author can hear the sound of a chainsaw as yet another skeleton of an erstwhile fine tree is removed.

Lightning

Lightning is a transient atmospheric electric discharge with a very high current, along a path of some 5–10 kilometres in length (figure 3.9). It is produced by the accumulation of electrical charges in the cumulonimbus clouds of a thunderstorm (or occasionally in snowstorms, dust storms, or in the clouds produced by volcanic eruptions). The atmosphere is normally electrically neutral in the troposphere (the lowest layer of the atmosphere), but it has been suggested that, as large raindrops, hailstones and ice pellets (together called hydrometeors) fall and collide with smaller water droplets and ice crystals suspended in the cloud, they transfer their electrical charges. The larger hydrometeors gain a negative charge as they fall, while the smaller hydrometeors are left behind the upper part of the cloud, with a positive charge. The cloud becomes positively charged at the top, and negatively charged below (Elsom, 1989), with a potential difference of up to 3 million volts per metre.

A typical cloud-to-ground lightning flash begins with the formation of an intermittent, highly branched discharge, called a *leader*, which moves downward with a velocity of about 100 kilometres per second. The leader carries a negative charge towards the ground along the tips of the branching discharge, and leaves a trail of ionised air. As a branch of the stepped leader comes to within about 100 metres of the ground, the negative charge in it attracts a positive charge just above the ground. The electrical field beneath the leader becomes very large, and one or more discharges form at the ground, creating *streamers* which rise until one (sometimes several) attach to the leader. When contact is made, the first bright return stroke occurs. This is the start of the lightning strike proper. The return stroke is a very large pulse of current that starts at the ground and runs back up the channel of ionised air into the cloud. It has a peak current of between 10 000 and 30 000 amperes, with a potential of 10 to

Figure 3.9 Worldwide, lightning is common, with about 100 lightning flashes every second.
Photograph courtesy Bureau of Meteorology, Australia.

200 million volts, and energy of perhaps a billion joules, a tiny proportion of which may be released as thunder. The process requires only a few milliseconds, and a single event can contain multiple repetitions — typically three or four complete strokes, but as many as 100 have been measured. As the human eye cannot resolve the time interval of a few thousandths of a second between the strokes, we perceive lightning as a flickering phenomenon (Reiter, 1992).

Since ancient times fantastic tales of 'ball lightning' have also been reported. Singer (1991) has reviewed the present state of knowledge based on objective observations and experiments which demonstrate that 'ball lightning' does occur and can be reproduced experimentally. The phenomenon is expressed as slowly moving reddish-orange spheres of plasma less than a metre in diameter which are actually able to penetrate a window or even a wall. The balls are rather unstable and frequently disappear with a frightful explosion.

The air through which a lightning flash passes is intensely heated — the electric current raises the temperature in the lightning channel to 25 000°C in a millisecond or less. The hot air expands suddenly, creating a shock wave with pressures of between 10 and 30 atmospheres. People within a metre or so of this shock wave can be thrown into the air, or suffer acoustic shock. Elsom (1989) reported a case of some tourists in Italy struck by lightning as they hid under a bush during a thunderstorm — the shock wave did not harm the victims, but shredded their clothes, and they emerged naked but, amazingly, not injured! Serious accidents have occurred, however, when people have been struck by lightning. Most deaths result from electric shock or burns from direct strike. Even if not fatal, the current can cause damage to the central nervous system, heart, and lungs (Andrews et al., 1992). Swimmers are especially at risk if lightning strikes the water nearby, although less so in salt water which is a better conductor than is fresh water, and more likely to convey the current around the person. Despite population increases, deaths from lightning strike have declined in technologically advanced countries over the last century (see figure 3.6), mainly as a result of fewer workers being employed in outdoor occupations.

Electronic circuits, including telecommunications facilities, are prone to damage if exposed to excess current or to current of wrong polarity. A lightning surge may enter a building by way of an unprotected power circuit, arc the system, and start a fire. Lightning strike may also be delivered to a person using a telephone via the telephone wire. The typical passenger aircraft in regular service will receive one strike per year, but as the airplane forms part of the lightning path, damages are rare, although sensitive electronic equipment may be rendered unusable by a strong electrical charge, with adverse effects on its controllability. If lightning strikes an object with moisture in it, the moisture can boil instantaneously. The sudden vaporisation can cause the object to expand explosively, whether it is a tree, a road, soil or a brick wall. Otherwise, the most damaging effects of lightning on the environment are as an occasional cause of wildfires.

Apart from prudent action in the event of a thunderstorm, management of the lightning hazard is mostly by installation of lightning conductors on vulnerable entities. A lightning conductor rod protects a building by readily setting up a discharge whenever a stepped leader approaches within 'striking distance'. The conductor is pointed so that it builds up a high density of positive charge in response to the negative charge of the leader above. Isolated trees and tall structures behave in the same way. The essential elements of an interception protection system are the air termination conductors to provide strike attachment points, the down-conductors to convey the lightning toward earth, and the earth electrodes to convey the lightning current into the earth (Andrews et al., 1992).

Hail

Hail is often associated with thunderstorms. A large hailstone exhibits roughly concentric accretions of clear and opaque ice, the nucleus of which is a raindrop carried aloft by updraughts and frozen. The successive accumulations of opaque ice (*rime*, formed by impact with supercooled water drops which freeze instantaneously) and clear ice (*glaze*, formed by freezing of a wet surface layer gained by passage through part of the cloud with a large liquid water content) occur as the hailstone is recycled through the cloud by successive entrainment in downdraughts and updraughts. The process is facilitated by the typical arrangement of downdraughts and updraughts in relation to the movement of the thunderstorm cell (see figure 3.8). Small (up to 6 mm dia.) hailstones consisting entirely of clear ice may result from the freezing of individual raindrops. Worldwide, average annual losses from hail damage to crops exceeds $2 billion, but high-value crops are not usually grown in hail-prone areas. There is no place where hail is so frequent and/or damaging that it precludes human activities. Occasional losses are accepted by many, and insurance may be a viable option.

Severe hail was recorded from the Sydney storm of 21 January 1991 (see figure 3.10), but earlier, on 18 March 1990, a hailstorm in Sydney produced large hail (dia. $\geq$ 2 cm) in a swathe at least 45 km long and up to 10 km wide, from west of Liverpool to the coast near Narrabeen. Hailstones 8 cm in diameter fell in one area, and hailstones over 5 cm in diameter were recorded from a number of localities. Insurance claims, mostly for hail damage, but also for wind damage associated with the storm, reached $313 000 000, making the storm the most damaging thunderstorm event in Australia to time of writing. About a year later, on 22 January 1991, a severe thunderstorm produced hail of unprecedented size over metropolitan Adelaide — hailstones with diameters of 10 cm were measured. Damages were estimated at $25 000 000, about half of this amount being for damage to motor vehicles (Watson, 1992).

TORNADOES

A tornado is a narrow swirling column of air extending from within the generating cumulonimbus cloud down to the ground: it usually appears as a dark, funnel-shaped cloud pendant from the parent cloud. Its most distinctive feature is the central core of greatly reduced pressure, in which air spirals upward at a very high velocity. A waterspout is simply a tornado formed over water — the distinctive appearance being due to water droplets sucked up into the central low pressure core (*not* liquid water, as is commonly believed). The term 'whirlwind' is a general term applied to

Figure 3.10 Hail damage to residential building in Duffy's Forest, Sydney, on 21 January 1991. Hailstones were of 'cricket ball' size.
Photograph David Nottage, Bureau of Meteorology. Reproduced by courtesy of Bureau of Meteorology, Australia.

any rotating column of air. When charged with dust or smoke particles, a whirlwind may become visible, and be termed a 'dust devil' or 'smoke devil'. The presence of a parent cloud, from which the rotating column of air descends, signifies a tornado or waterspout.

Most of the world's tornadoes occur in the United States, although a considerable number are reported from some other countries, including Australia, and even the UK. The development of tornadoes requires severe weather systems, such as violent thunderstorms, in which air masses contain high levels of moisture with large amounts of latent heat energy. Considering the short life of a tornado (for any given place, probably not more than a minute or so), the violence of its nature, and the spontaneity of its formation, it is understandable that the behaviour of tornadoes is not well understood. However, it is known that the formation of a tornado requires a surface layer of moist warm unstable air, a mechanism that allows cool drier air to overrun the warm surface layer, and a well developed flow of strong winds between around 3000 to 6000 metres. Conditions conducive to tornadoes can occur when a cold front moves so rapidly aloft that the surface front temporarily entraps warm moist air underneath; or when the lower leading edge of a severe thunderstorm, formed by cold air rushing out of the storm, overruns the warm surface layer which is being drawn toward the rising core. In the latter case, the inflow of moist warm air beneath the front of the storm is in the opposite direction to the outflow of cold air above it, leading to friction or wind shear and the development of vorticity.

The destruction caused by tornadoes is the result of both extremely high rotary wind velocity and the very low pressure within the vortex. Estimates based on engineering studies of tornado damage indicate that horizontal wind speeds surrounding the vortex may be as high as 500 km/h. The low pressure inside the vortex can only be estimated, but it is possible that the pressure may drop to as low as 800 millibars (i.e., about 80% of normal atmospheric pressure at sea level) in some tornadoes. Snyder (1991) compares the forces exerted on the outside of a building by a tornado to those responsible for keeping an intercontinental jet aircraft in the air. Although impact is severe in the path of a tornado, that path is quite narrow, usually only a few hundred metres wide. Most tornadoes impact the ground for a few tens of kilometres at most — the record is 352 km. Tornado impacts are principally due to the direct force of wind (see discussion above), but the extreme conditions associated with tornadoes have resulted in quite large objects becoming airborne, which in turn may act as missiles (figure 3.11). Even people become airborne, and Carter et al. (1989) concluded that most deaths and many serious injuries to people in tornadoes resulted from that cause.

Management of the tornado hazard by modification of the tornado itself is not feasible, and ability to forecast is poor. Due to the very narrow path of an individual tornado, and the unpredictability of its passage at the local scale, general evacuation from an area at risk may simply result in placing more people in the path of the storm, in vulnerable motor vehicles. Downward evacuation, or retreat into a strongly constructed basement, is the most feasible option, and may be undertaken with very short warning time. Although it is possible to design buildings to be resistant to winds of tornado strength, it is doubtful whether it is cost-effective to build them, as even in tornado-prone areas, probability of a direct hit by a tornado on an individual structure is very low. Compliance with standards for extreme non-tornadic winds, in terms of strength requirements for fixed structures, and tie-down requirements for mobile homes, seem to be the most appropriate measures, coupled with appropriate loss-sharing arrangements, as through insurance.

TEMPERATURE EXTREMES

Quite apart from the effect of desiccation, a few days of extreme temperatures can significantly reduce the yield of some crops. Humans, moreover, are essentially warm climate animals, and must keep inner

Figure 3.11 This small sportscar became airborne in a tornado in Baldwin, Mississippi, USA, lodged in a tree, and broke up, before the tree itself was uprooted by the tornado.
Photograph courtesy American Red Cross.

body temperature within a very narrow range, or irreparable harm ensues. Naked and without shelter, most of us are comfortable over the range of humidities normally experienced in the atmosphere if temperature is between about 20°C and 25°C, but become increasingly uncomfortable if the temperature moves markedly higher or lower, with death possible at either extreme. The body gains or loses heat to its surroundings by radiation, convection, and conduction, gains heat by its own metabolism, and loses heat by evaporation (sweating and breathing).

High temperatures

When temperature rises, the temperature perceived by a normal person is a function of both air temperature and relative humidity. Where relative humidity is high, apparent temperature will be above actual air temperature, and vice versa (see figure 3.12): the average person, suitably clad, in the shade and subject to mild air movement (around 10 km/h), at an ambient temperature of 38°C and relative humidity (RH) of 20 per cent, will actually *perceive* a slightly lower temperature, but will perceive a higher temperature if RH is raised above 25 per cent. The degree of heat stress

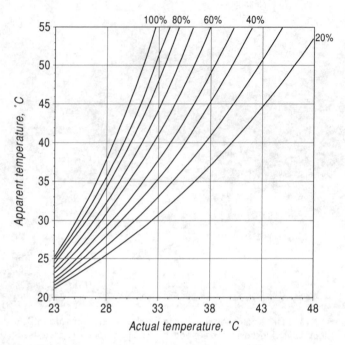

Figure 3.12 Relationship of apparent (or 'effective') temperature to actual air temperature and relative humidity.
Data from Steadman (1979).

varies with an individual's physiology, age, and health (Steadman, 1979), but objective factors contributing to heat stress are:

- lack of air movement — even a slight breeze usually aids in reducing heat stress, especially in the lower ranges of relative humidity, since the principal cooling agent for the body is the evaporation of sweat into the atmosphere;
- presence of air temperature greater than body temperature, resulting in net gain in body heat from convection;
- exposure to radiant heat, especially incoming solar radiation, although sources of high infrared radiation, such as bright incandescent lights, cooking devices, or hot industrial equipment may be implicated;
- insufficient supply of fluids to reduce perspiration loss;
- physical activity;
- inappropriate clothing — absorbent, light-coloured fabrics are best in high temperatures; and,
- high basal metabolic heat production — largely a function of diet.

Extreme temperatures in excess of 50°C (122°F) have been experienced in numerous places in all states of Australia, except Tasmania, where extremes have exceeded 40°C (104°F). From time to time, heat-waves, or extended periods of elevated air temperature, result in great discomfort to most people and many animals, and in death to some. Hosts of birds and other animals perished in southern and central Australia in 1932, when for over two months the air temperature remained above 38°C (Lack, 1954). In humans, actual certification of death due to excess heat as immediate cause of death is rare, but Oeschli and Buechley (1970) and Clarke and Bach (1971) found that daily numbers of deaths reported in a heatwave increase both with increasing temperature and advancing age of victim, and that the time lag between temperature maximum and mortality maximum is about 24 hours. City dwellers are most at risk, because of the urban heat-island effect (Buechley et al., 1972).

Low temperatures

The hazard of *hypothermia*, or sub-normal body temperature, occurs when the core body temperature falls more than about 2°C (3.6°F) below the normal level of 37°C (98.6°F). If the state lasts for several hours, death may occur. If a person's core body temperature falls below about 32°C (90°F) at any time, there is a significant chance of death. Older people are especially endangered because their bodies cannot regenerate heat as quickly as those of younger individuals, and the sensory mechanism which detects a drop in body temperatures is also less efficient in elderly people. Unaware of their danger, older people may die without warning. Furthermore, heat loss can be made worse when wearing wet clothes because water conducts heat much more efficiently than air. A person caught in wet clothes, as may happen in a snowstorm, can develop potentially fatal hypothermia at temperatures above freezing point, especially if there is a high wind velocity. It is the combination of high wind velocity and low temperature that usually creates the hazard of hypothermia, since the rate of cooling of uncovered skin, at any given temperature below freezing, depends on the wind velocity. When the ambient air temperature is –10°C, and the wind blows at 75 km/h (approx. 40 knots), the rate of body heat loss is equivalent to the rate at –25°C under calm conditions. The equivalent temperature experienced by a person exposed to both extreme cold and wind is known as the *wind chill equivalent temperature* (figure 3.13). Important factors which also influence the impact on a person, and which may be critical to onset of hypothermia, include: solar radiation input; insulating effect of clothing and its colour (dark clothing absorbs solar energy more effectively), and its moisture content (cold stress can occur at temperatures well above freezing if clothing is saturated); amount of physical activity being undertaken; and basal metabolic heat production (largely a function of diet).

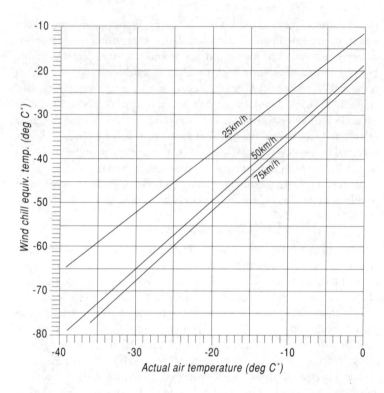

Figure 3.13 Wind chill equivalent temperatures at wind speeds of 25, 50, and 75 kilometres per hour.
Data from Steadman (1971) and Schlatter (1981).

Blizzards, freezing rain, fog and frost

Snow occurs when the freezing level in the atmosphere is so near the surface (usually <300 m) that there is not time for the assemblages of ice
crystals to melt before reaching the ground. Snow is rare when surface air
temperature exceeds 4°C. Mixed snow and rain or *sleet*, is especially likely when air temperature at the surface is around 1.5°C. Snow creates a
particular hazard when it falls heavily in an area which does not customarily receive snow, especially urban areas. Where specialised equipment
for removing the snow is not available, chaos may ensue.

Blizzards

A blizzard is a strong wind laden with powdery snow. The snow may be
either partially or entirely picked up from the ground. Visibility in many
blizzards is reduced to zero. While the total amount of snowfall may be
less than that of a regular winter snowstorm, the blizzard may produce

hu ge, dune-like snowdrifts which paralyse traffic and may trap people in their cars. At what point a snowstorm is called a blizzard is somewhat arbitrary: blizzard-like conditions may momentarily occur with almost any severe winter storm. In the USA, the Weather Service issues blizzard warnings when wind speeds of at least 56 km/h (30 knots) combine with considerable amounts of falling or blowing snow, and visibility is dangerously restricted. The hazards associated with blizzards (low temperatures, wind chill, low visibility, insurmountable snow drifts) can occur in all terrains. Consequently, blizzard conditions have not only trapped skiers, mountain climbers, and explorers, but have endangered people in cities, and on the plains. Sometimes people in cars engulfed by snow have attempted to remain warm by keeping the heater on and engine running, unaware that deadly carbon monoxide produced by the engine was also being trapped by the blanket of snow.

Freezing rain

Glaze, or ice formed from supercooled raindrops freezing on impact, may cause great damage to trees, power lines and power poles, as the affected entities simply collapse under the vastly increased load imposed by the ice. While flying through fog or clouds at below-freezing temperatures, aircraft can also accumulate glaze: cloud droplets, striking the leading edges of the aircraft, freeze on impact. This clear ice can build up rapidly and is quite difficult to remove. Another type of ice deposit may form on an aircraft when it flies through a cloud composed of supercooled droplets which are too small to precipitate: these droplets, on striking the plane, build up rough, irregularly shaped ice, or rime. The accumulation of ice, either glaze or rime, can radically change the aerodynamic characteristics of an aircraft in several ways: lift is decreased; friction, or drag, is increased; and total mass is also increased. These effects may combine to decrease speed and manoeuvrability of the aircraft to a level where it stalls and/or becomes unmanageable.

Fog

Condensation in the form of fog is often produced as a result of the cooling of a warm, moist parcel of air which moves over a cold land surface. Hence, fog in coastal areas often results from a moist air mass from the sea moving over the relatively colder land in cold weather. Conversely, 'steam fog' may be caused by evaporation into cold air of moisture from a warm water surface. On a clear winter night, radiation cooling of the surface of the earth can result in the cooling of the lower air layer to the point when condensation in the form of fog occurs, either *in situ* or in valley bottoms where the cold air accumulates from cold air drainage. Fog presents a serious hazard to all types of visual navigation, especially to motorists, but even the pilot of a radar-guided aircraft must be able to see in order to land it. Cold fogs have been successfully dissipated at a local scale by the dispersion of tiny particles of frozen carbon dioxide ('dry ice'), or of propane gas, to stimulate freezing and subsequent rain-out of the ice particles. Some limited success has been achieved with warm fogs by artificially heating the air, or using high-powered fans to draw down dry air from aloft.

Frost

Hoar frost, which is a veneer of ice crystals, can form when the surfaces of objects are at 0°C or below, the surrounding air is saturated at 0°C, or slightly below, and there are nuclei present for the process of sublimation (when water passes directly from gaseous phase to solid) to occur. Nuclei are as essential for ice crystal formation as they are for cloud droplet formation, but are always present on plants in the form of dust particles, plant hairs, irregularities of plant parts, etc. A *glaze* form of frost may also occur if precipitation of dew is followed by decline in temperature below freezing — the dew droplets then solidify into an amorphous ice covering on objects. The phenomenon of *frost-heave*, disruption of the ground surface by freezing of water in the soil, or rupture of water pipes, is not strictly a form of frost, neither is *frostbite*, which is simply due to the freezing of skin or tissues of the extremities of a person or animal, rupturing cells and killing the tissue (Killian, 1980).

The frost hazard to crops may be associated with freezing air temperatures due to cold air drainage, or to radiation, which occurs on calm, clear nights when outgoing terrestrial radiation is unimpeded by cloud or fog, a condition sometimes accompanied by temperature inversions, as air near the ground becomes chilled and heavy. At a dew-point temperature of $\leq 6°C$, frost is probable. Presence of moisture aloft, even in the absence of cloud, will impede outgoing radiation. Arrival of an air mass of significantly higher pressure during the night increases chance of frost. The actual temperature at which damage occurs depends both on the type of plant and its stage of development. Some plants cease functioning at temperatures either higher or lower than 0°C — thus a plant may suffer 'chilling' injury quite apart from effects of hoar frost or the existence of freezing temperatures. Direct physical damage to plant tissues from frost is caused by the water that is part of plant cells solidifying and expanding, followed by the bursting of cell walls and the deterioration or death of plant tissue.

Apart from frost insurance, usually available where probability of frost is low, most effort in frost management has been directed to modification of either the plant or of its microclimate (Berg and Wright, 1984). It is hard to imagine a farmer who would plant sensitive crops if they were virtually certain to be exposed to frost, so probably the strategy of hazard avoidance should also be included. This may be as sophisticated as using microclimatic zonation to identify areas of potential freedom from frost using detailed terrain analysis, aided perhaps by thermal remote sensing. Plant scientists have also been active in the development of frost resistant varieties of crops which mature faster than native types, allowing growth to maturity in the (sometimes brief) frost-free season. There are also techniques for frost-hardening plants and for regulation of bud development.

Temporary modification of the microclimate of susceptible crops (usually aimed at elevating the temperature in the vicinity of the crop by a few degrees) is attempted by techniques aimed at increasing the temperature of surface air by mixing it with higher and warmer layers; directly heating the surface air, using heaters, or sprinklers to access the latent heat released in condensation or sublimation (as water freezes, it releases latent heat); and reducing outward radiation by creating an artificial fog or smoke cover or by using plant shields (Bagdonas et al., 1978). Dynamic frost mitigation measures are usually initiated on the basis of short-term forecast, and often in association with manual or electronic monitoring of key areas on an individual farmer's property.

CONCLUSION

A still atmosphere is a dead atmosphere — the restlessness of the atmosphere is vital for life on earth. The same weather patterns that are occasionally responsible for disasters are life-giving most of the time. Tropical cyclones are to be feared, certainly, but much of the tropical and subtropical world derives a significant proportion of its annual rainfall from these storms. Thunderstorms usually bring welcome rainfall, and not infrequently a welcome cool change to the same people who very rarely suffer wind or hail damage, or, most unlikely, lightning strike. We have altered the behaviour of the atmosphere unintentionally, by the increase of 'greenhouse' gases, by release of industrial particulates which have changed rainfall patterns, and through the urban 'heat island' effect, but intentional modification of the atmosphere has not proven to be unequivocally beneficial, nor does it promise to help reduce atmospheric hazards. Cloud seeding (with tiny crystals of frozen carbon dioxide or silver iodide), to encourage rainfall from suitable clouds, for example, has been claimed by its proponents to be a demonstrated way of controlling precipitation, and there is evidence that heavy seeding of a tropical cyclone at frequent intervals, say every couple of hours, can help reduce its central pressure. The reduction in intensity may, however, involve changes in the radius of strong winds, and in the behaviour of storm surge, thereby affecting areas which may have been safe if the storm were left alone. Forcing extra rainfall over sea may mean less of the desirable rainfall over land, and who can say if the path of the storm has been altered by human interference? Seeding of every potentially hazardous tropical cyclone would involve enormous resources, and where the outcomes cannot be shown to be positive, who will pay?

Chaos theory has indicated that long-range forecasting of atmospheric behaviour will probably always remain beyond our grasp, and indeed even quite crude models of the atmosphere require the most powerful computers in the world on which to run. We stand humbled before the very air we breathe, and the best defence against its occasionally violent behaviour is to be prepared: prepared to modify our behaviour; prepared to modify our land-use patterns; and prepared to incorporate suitable hazard-mitigation measures into the structures in which we live and work. And having done all, prepared to suffer the rare catastrophic loss against which no form of protection can be economically feasible. It is Nature's rent.

■

4

Earthquakes, volcanoes and mass movement — Terra non firma

The complacency with which people inhabit land near, or even on, tectonic plate margins is based on the fact that the surface of the earth is, in general, sufficiently stable for dwellers thereon to ignore its potential for rapid change. But the complacency is periodically shattered by earthquakes and volcanic eruptions, possibly the most feared of all natural disasters, and at the local scale by sudden failures of surface stability in the form of landslides, lahars, or avalanches. Some of the great earthquakes of history have produced disaster tolls which almost defy comprehension: at Shensi, China in 1556, when almost a million people perished, or at Calcutta in 1737 when 300 000 died. In the twentieth century over 1100 fatal earthquakes have affected seventy countries and killed over 1.5 million people. Most of the crust of the earth is of magmatic origin, evidence of the cardinal role of volcanic and related magmatic processes in forming the outermost solid skin of our planet. About 50 volcanoes erupt each year (Simkin et al., 1981). Individual volcanoes may remain inactive for centuries or even millennia, and be considered dormant or even extinct, but some of the worst volcanic disasters in history have been produced by volcanoes believed to be 'extinct', for example, Vesuvius, that destroyed Pompeii and Herculaneum in 79 AD (Sigurdsson et al., 1985), or Mount Lamington in Papua–New Guinea in 1951 (Taylor, 1958).

EARTHQUAKES

The ground is made to vibrate (or 'quake') almost continuously, but only to a very minor extent, by disturbances such as the wind shaking trees, or traffic movements. Occasionally, amid this background of subdued random ground motion, there occurs a burst of more intense vibration in which different groups of waves arrive at a surface position in a regular order. A seismic event of this type is called an earthquake. Most are detectable only by instruments, but some are powerful enough to harm people and to damage structures.

Seismic (earthquake) waves are generated as the result of a abrupt release of energy when rocks which have been strained elastically suddenly fail and move (see figure 4.1). The point of rupture, from which the seismic waves are transmitted in widening circles, is known as the *focus*. It can be between the earth's surface and depths of some 600–700 kilometres. Shallow focus earthquakes (i.e., those less than 40 km below the surface) are the most damaging events. The *epicentre* is that point on the

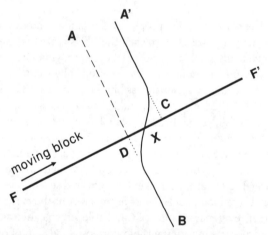

Figure 4.1 Earthquake mechanism. Slow differential movement of the blocks on one or both sides of a fault occurs, but this initially does not cause slip along the fault plane, which is locked by friction. The diagram shows a fault F–F' along which horizontal displacement is slowly taking place, causing an originally straight line A-B across the fault to become deformed with movement into A'–X–B. Eventually the elastic distortion and the stresses generated close to the fault are sufficient to overcome friction along the fault plane, and movement takes place suddenly, from X to C on one side, and from X to D on the other, to produce two discontinuous lines A'C and DB, displaced across the fault. The corresponding release of stored energy generates an earthquake.

earth's surface directly above the focus, and is the source point for earth-quake measurement. Although some earthquakes are caused by volcanic eruptions, the commonest sources of shallow earthquakes are faults, which are frequently (but not exclusively) associated with active tectonic plate margins. The principal outcome of earthquakes is ground shaking, which can be decomposed into four types of earth movement. The P wave, or primary wave, from earthquakes is a *compressional* wave which resembles the effect of a shunt on a line of connected railway trucks. It spreads out from the focus at a velocity of around 8 kilometres per second, depending on the density and elastic properties of the rock through which it travels. P waves are able to travel through both solid rock and liquids such as the ocean. S or secondary waves move at about half the velocity of primary waves and cause *vertical* ground shaking. S waves cannot propagate in the liquid parts of the earth but when they reach the surface the resulting ground motion is highly damaging to structures. *Love waves,* which are surface waves, create a particular prob-lem with regard to building failure, since the ground surface is shaken *horizontally* at right angles to the direction of propagation of the wave, thus creating horizontal stress on structures which were designed for ver-tical stress. *Rayleigh waves* (also surface waves) operate rather like ocean waves, except that the orbital motion is reversed with respect to the direction of propagation of the wave.

The severity of ground shaking at any point depends on a combination of the magnitude of the earthquake, the distance from the focus, and local geological conditions which may either amplify or reduce the earth-quake waves. Acceleration of the earth's surface, or the rate at which the earth is moved, is measured in units of one g, or the acceleration due to gravity. Acceleration greater than 1 g in the vertical plane would result in unsecured objects leaving the ground. Peak accelerations during severe earthquakes may exceed 2 g. The greatest structural damage is created by horizontal ground movement. Buildings are constructed to resist the pull of gravity and can therefore withstand some vertical movement. How-ever, weak structures may be unable to cope with horizontal ground accelerations as little as 0.1 g.

Some earth tremors are caused by people. Extraction of coal may cause collapse of the strata above the seam into the abandoned workings, pro-ducing small seismic foci; and the impounding of great volumes of water in very large reservoirs may result in uneven and sporadic settlement under the new load, and lubrication of pre-existing faults by water under pressure (Gupta, 1992). The Koyna (India) Dam was constructed in 1962 in a very non-seismic area. After filling, a definite correlation was observed between the height of the water in the reservoir and the seismic activity, terminating in a major earthquake in 1967. In the vicinity of Denver, Colorado, a few years ago, a disposal well for waste fluids was drilled to a depth of 3700 metres into Precambrian crystalline rocks. Subsequently, a very striking relationship (correlation, $r = 0.652$) was observed between the volume of fluid injected and numbers of earthquakes, in an area which had previously been earthquake free.

Measurement of earthquakes

While the vibrational energy released by an earthquake may be measured on an absolute scale, the effects of the earthquake on cultural entities depend on a variety of factors, principally distance from earthquake epicentre and ground conditions. There are therefore two ways to measure earthquakes: (1) *magnitude,* which is related to the vibrational energy of the shock — a given earthquake has a single, precise value of magnitude associated with it, and (2) *intensity,* which is a measure of the damages to human life and property as a result of the shock, and varies from place to place. The most commonly used *intensity* scale is the Modified Mercalli (or MM) scale, shown in table 4.1, and the standard scale for measurement of earthquake *magnitude* is the Richter surface-wave formula:

$$M_S = \log_{10} (A/T) + 1.66 \log_{10} Q + 3.30$$

where A is the amplitude of ground motion (in microns);
T is the corresponding period (in seconds); and
Q is a correction factor that is a function of distance (degrees), between earthquake epicentre and recording station.

Illustrative values of Richter magnitude are shown in table 4.2.

The impact of earthquakes

It is an earthquake engineers' maxim that 'earthquakes don't kill people, buildings do', and most earthquake impact is due to ground vibrations (cf. figure 4.2). During earthquakes the ground may vibrate at frequencies ranging from 0.1 to 30 Hertz. High-frequency waves tend to have high acceleration but relatively small amplitude whereas low frequency waves have small acceleration but large velocity and displacement. *P* and *S* waves (defined above) are mainly responsible for the high frequency vibrations (greater than 1 Hertz) which are most effective in shaking low buildings. Rayleigh and Love waves are of lower frequency and are usually more effective in causing tall buildings to vibrate. Consequently, the

Table 4.1	Modified Mercalli scale of earthquake intensity	
Intensity	Impact	Effect
I	Negligible	Detected by instruments only.
II	Feeble	Felt by sensitive people. Suspended objects swing.
III	Slight	Vibration like passing truck. Standing cars may rock.
IV	Moderate	Felt indoors. Some sleepers awakened. Sensation like a heavy truck striking building. Windows and dishes rattle. Standing cars rock.
V	Rather strong	Felt by most people; many awakened. Some plaster falls. Dishes and windows broken. Pendulum clocks may stop.
VI	Strong	Felt by all; many are frightened. Masonry chimneys topple. Furniture moves.
VII	Very strong	Alarm; most people run outdoors. Weak structures damaged moderately. Felt in moving cars.
VIII	Destructive	General alarm; everyone runs outdoors. Weak structures severely damaged; slight damage to strong structures. Heavy furniture and monuments toppled.
IX	Ruinous	Panic. Total destruction of weak structures; considerable damage to specially designed structures. Foundations damaged. Ground fissured.
X	Disastrous	Panic. Only the best buildings survive. Foundations ruined. Rails bent. Ground badly cracked. Large landslides.
XI	V. disastrous	Panic. Few masonry structures remain standing. Broad fissures in ground.
XII	Catastrophic	Mega-panic. Total destruction. Waves are seen on the ground. Objects are thrown into the air.

effects of ground shaking will vary greatly depending on the wave composition of the earthquake and the resonant frequency of structures. For example, in the 1985 Mexico earthquake some of the taller buildings survived because their natural resonant frequency did not match the high frequency of the shock waves, while many shorter buildings collapsed. Conversely, the 1964 Alaskan earthquake, which produced mainly low frequency vibrations, proved relatively ineffective in toppling low-rise structures, the popular building type there.

Ground motions in unconsolidated material are enhanced in both amplitude and duration compared to those recorded in solid rock. Consequently, buildings located on floodplains and ancient sedimentary deposits such as lake beds are more vulnerable than those on solid rock. Close to its source, the Mexico earthquake of 1985 (Richter magnitude, $M_S = 8.1$), did little damage relative to its great energy. The enhancement of seismic waves in limited parts of Mexico City (at a distance of 420 kilometres from the source!) contributed to most of the losses from

Table 4.2   Richter earthquake magnitudes

0	Smallest detectable.
2.5	Felt by humans if nearby. About 100 000 each year $\geq M_s$ 2.5.
4.5	Capable of causing local damage.
5	Early atomic bombs were comparable — energy release about 2×10^{12} Joules.
5.6	Newcastle (NSW) earthquake of 1989 — energy release about 1.56×10^{13} Joules.
6	May be locally destructive. About 200 per year $\geq M_s$ 6.0. About 20 earthquakes $\geq M_s$ 6.0 have occurred in Australia in the twentieth century.
6.4	Comparable to one megaton Cold War nuclear missile.
6.8	Meckering (WA) earthquake of 1968 — energy release about 1×10^{15} Joules.
7	Major earthquake, recorded over entire earth. About 15 per year.
8	About one per year.
8.9	Maximum recorded.

the event. Acceleration records on solid rock showed values of around 0.04 g, compared with observations taken in central Mexico City, which is founded on a dried lake bed, where measured peak accelerations reached 0.2 g. Although local building codes had already incorporated a factor to allow for the soft substrate, the amplification and duration of the ground shaking in parts of Mexico City were greater than had been anticipated, and a number of 10–24 storey reinforced concrete structures collapsed (8000 died).

Perhaps the most serious hazard associated with soft sediment is soil liquefaction, which occurs when water-saturated sediments, because of strong shaking, temporarily lose their strength and become fluid. Mass movement may occur on slopes of less than 3° when liquefaction occurs. Material may be displaced by tens of kilometres at velocities of up to tens of kilometres per hour. There may also be horizontal displacement of surface blocks as a result of liquefaction of a subsurface layer. If liquefaction occurs at some depth but slopes are too gentle to allow lateral displacement, ground oscillation commonly occurs. This is often observed as a travelling ground wave accompanied by the opening and closing of fissures. In the 1964 Alaskan earthquake, cracks up to 1 metre wide and 10 metres deep were observed under such conditions. Loss of bearing strength occurs when the soil liquefies under a building. Large deformation may result within the soil body, causing structures to settle into the ground surface and/or to tip.

The severe shaking in an earthquake may result in natural slopes weakening and failing. The resulting landslides, rock or snow avalanches

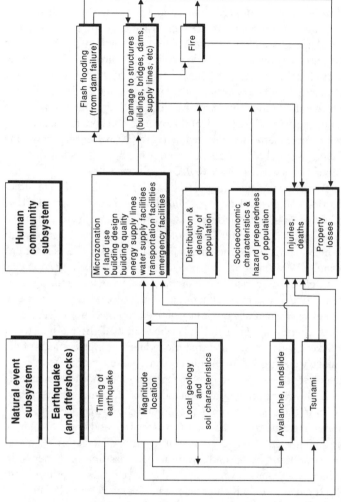

Figure 4.2

Interrelationship of factors in earthquake hazard. The primary outcome of an earthquake is ground displacement, but this in turn sets up secondary earthquake phenomena, such as ground shaking, and differential ground settlement; soil liquefaction; landslides and mud slides (incl. submarine avalanches or debris flows); snow and ice avalanches; tsunami and seiches; floods from dam failure or release of lake waters, and creation of new lakes from ground displacement or landslide. In urban areas fire and explosion following the earthquake may cause damage which exceeds that from the direct effects of ground shaking.

are major secondary contributors to earthquake disasters. Rock falls are probably the most common earthquake-induced form of slope failure but the two leading causes of death are rock avalanches and rapid soil flows. Rock avalanches are often of large volume ($\geq$ 1 million cubic metres of material) and may travel several kilometres from source at velocities of up to hundreds of kilometres per hour. The greatest landslide disaster ever recorded, a rock and snow avalanche from the overhanging face of Nevados Huascaran mountain in Peru, was due to an earthquake offshore of Richter magnitude 7.7. A turbulent flow of mud and boulders in excess of 50 million cubic metres was released in a surge 30 metres high, travelling at an average speed of over 250 kilometres per hour. The flow buried a number of towns where at least 18 000 people lived.

The economic impacts of an earthquake are not limited to fatalities and injuries, and physical damage to buildings and other structures. Loss of production in industry and services, unemployment, increased public health and social expenditures, and other secondary 'ripple' effects can greatly increase the total economic impact of an earthquake. Kawashima and Kanoh (1990) present a method for calculating the indirect effects (negative effects on agriculture, fisheries, mining, manufacturing, construction, utilities, commerce, finance and insurance, transportation, and service industries, partially offset by positive economic impact generated by reconstruction activities). The authors applied the method to an earthquake in Japan, and found that the indirect, or 'ripple' effects of the earthquake were larger than the direct damage by a factor of 2.5.

The distribution of earthquakes is far from random. Most are closely related to the distribution of tectonic plate margins, and a large proportion occur in the so-called 'rim of fire' around the Pacific. Hewitt (1984) also draws attention to the fact that many occur in and at the margins of mountain lands and there is a concentration of damages in mountain foot, intermontane basin, and mountainous coastline settlements. In Australia, several quite severe earthquakes have occurred far from centres of settlement, with a Richter magnitude 6.8 event severely damaging the small Western Australian town of Meckering in 1968. However, in 1989 a modest (Richter magnitude 5.6) earthquake at Newcastle not only illustrated the potential vulnerability of urban areas in Australia to a serious earthquake, but clearly demonstrated the significance of underlying geology in controlling the effect of an earthquake on the built environment. The earthquake caused considerable damage locally, and was felt over a wide area, including Sydney and Canberra. It has been estimated that approximately 25 000 houses and 1000 commercial and public buildings suffered damage. Damages were of the order of $750 million (Berz, 1991). There were only twelve fatalities, an outcome partly of the time of day and time of year.

It is well known that earthquake intensity is greater where settlements are founded upon unconsolidated sediments, such as old lake beds or alluvia, and in the Newcastle earthquake there was a strong correlation between the areas of maximum ground motion, where intensity was up to MMVII Mercalli, and areas of relatively deep alluvium. Outside of these areas, apart from a few isolated pockets, the maximum intensities do not appear to have exceeded MMVI (Walker, 1990). Almost all major damage to buildings was associated with unreinforced masonry construction. Minor damage was also common to internal linings of buildings. Many unreinforced masonry structures varying in size from houses to multi-storeyed apartment and commercial buildings did perform well, however, even when founded on alluvium. Losses arising from damage to commercial and public buildings were much greater than losses arising from damage to houses, principally because the central business district is located in the alluvial area that experienced the maximum intensity of ground motion. Only a handful of structures suffered major structural collapse (collapse of floors or roof). Most damage was in the form of parapet and gable end failures; corner failures; transverse panel failures; concrete column failures, and failures consequent upon these factors. The most dramatic collapse was that of the relatively modern Workers Club building which led to the death of 8 people. Reports suggest that collapse of the roof may have been initiated by failure of a supporting unreinforced masonry wall, and that the floors collapsed as a consequence (Walker, 1990). No significant damage appeared to have occurred to the main structure of steel and concrete buildings in which the frame was designed to resist torsional motion, and very little significant damage occurred to timber buildings, owing to their inherent flexibility. Tiedmann (1990:45) concluded that quality of material and workmanship in brick buildings, or with walls made of brick, was one of the important factors contributing to damage, and went on to point out that 'bad workmanship should be considered a crime in seismic regions'.

Management of the earthquake hazard

Apart from efficient emergency and communication services, needed in all disasters, earthquake management principally revolves around the zonation of areas for earthquake risk, and the adoption of building codes consistent with the degree of exposure to risk. Zonation may be at the macro or micro scale — for example, all of Australia is zoned on a macro scale, with regions classified into one of four seismic zones, according to an assessment of their earthquake risk. Microzonation (cf. French, 1984; Carrara, 1991) uses a geographic and probabilistic approach to assessment of seismic risk (plus secondary hazards of liquefaction and landsliding), and involves detailed study of the substrata of an area, including evaluation of their potential reaction to ground shaking, as well as study of seismic records and geologic features which would indicate potential rupture sites. Geographic Information Systems (GIS) are normally employed (cf. Emmi and Horton, 1993), and satellite imagery can provide detailed information on surface features such as faults and lineaments which are important indicators of lines of weakness. At the time of writing, a project for the earthquake zonation mapping of major urban areas in Australia, sponsored by Emergency Management Australia, had commenced.

There are three major difficulties associated with the reduction of seismic risk. First, the understanding of earthquake generation is far from complete. Prediction of seismic risk (i.e., in a probabilistic sense) can be assessed from general theory and from past records of an area, and may be used for land-use planning and for laying down building codes, but prediction of occurrence and prediction of patterns of ground shaking remain difficult (Bolt, 1991; Peruzza and Slejko, 1993). Prediction of a specific earthquake is of value only if it allows evacuation of a threatened city and deployment of emergency services before the event. In some countries there is a well-developed folklore about animals sensing an oncoming earthquake, and Deshpande (1987) gives a measure of credence to some of the stories: some animals appear to be sensitive to slight changes in the earth's magnetic field or electric field, and it is well known that there are animals which can hear in the infra-sound or ultra-sound areas. There is a real-time earthquake monitoring system for the Bullet Trains in Japan, which can enable trains to be stopped immediately the earthquake occurs if track damage is likely (Nakamura and Tucker, 1988).

Second, competing social forces continue to prevent the optimum growth of, and application of, knowledge for earthquake hazard mitigation. The Armenian earthquake in December 1988 ($M_S = 6.9$, plus aftershock of 5.8 four minutes later) resulted in over 27 000 deaths, 100 000 serious injuries, and half a million people being made homeless (figure 4.3). It caused economic losses of over A\$32 billion. A majority of the newer buildings were not of seismically resistant construction even though the area had been known from antiquity to be geologically unstable. The problem was not that available expertise was lacking, but that earthquake risk reduction was afforded a very low priority within the

Figure 4.3 In Spitak and Leninakan, the cities most affected by the Armenian earthquake of 1988, destruction was almost total.
Photograph of collapsed stone masonry building in Spitak by H. Lew. Reproduced by courtesy of Earthquake Engineering Research Institute (from *Armenia Earthquake Reconnaissance Report*, suppl. to *Earthquake Spectra*, August 1989).

complex social, economic and political forces in that country, which was part of the now defunct Soviet Union at the time. The result was a catastrophic number of fatalities and economic dislocation: reconstruction was also seriously impeded by political factors, especially by the fact that planning and construction decisions were made by central authority far from the country concerned (Borunov et al., 1991).

Third, because of indecision between minimising loss of life and maximising broader benefits, general agreement on acceptable earthquake risk remains confused. Even if probability models are worked out accurately and are clearly explained, there still remains the difficulty of lack of agreement on the major goals of hazard mitigation. Improvement of the building stock may be seen to be to the detriment of the social environment (Geipel, 1991), although there has recently been a renewed interest in traditional, low-cost earthquake resistant construction techniques (Touliatos, 1993). In Western countries the Christian ethic has in the past put emphasis on the maximisation of life safety rather than on the minimisation of economic loss and indeed, in the USA, the uniform building code specifically states that 'the purpose of this code is to provide minimum standards to safeguard life or limb, health, property and public welfare while regulating and controlling design and construction'. Uncertainties involved in the estimation of earthquake stress on structures are usually allowed for by the application of safety factors: larger than expected ground motions, given a realistic lifetime of the structures, are usually adopted for design purposes. Such conservative judgments would generally be applauded, but there are now social demands for cost-effectiveness, and limits on construction costs.

Earthquake insurance is regarded as an important component in reducing both hazards before and losses after earthquakes. The greatest overall benefit from earthquake insurance accrues when the availability of low-cost insurance is linked to a requirement to upgrade the seismic resistance of the structure. For domestic dwellings, inspections at the time of purchase could establish premium levels according to the degree of risk inherent in the dwelling and its location. It has been suggested that home mortgage lenders should require earthquake insurance as a condition for granting a loan. Large earthquakes are very infrequent however, and the usual actuarial procedures used to predict probable losses and to spread risks (as in casualty and fire loss) are not very reliable (Bolt, 1991).

The relationship between fluid injection and earthquakes observed near Denver (noted above) has caused some engineers to speculate that, rather than letting strain accumulate on a fault until it is released in a major destructive earthquake, it may be possible by fluid injection to reduce the frictional resistance to faulting, and release the strain with a number of smaller shocks. However, because of the logarithmic relationship between the magnitude of an earthquake and the amount of seismic energy released, a huge number of shocks of small magnitude would have to be triggered, with uncertain impacts, to release the energy equivalent of an earthquake of large magnitude.

VOLCANOES

Throughout human history, volcanic eruptions have engendered fear, superstition, fascination, scientific curiosity, and, in more recent times, a rational approach to the planning of land use and human activity in their vicinity. The total number of volcanoes on the planet is probably in the tens of thousands (there are at least 10 000 in the Pacific region alone), but most are either dormant or extinct. There are about 500 active volcanoes, with a concentration in the Pacific 'ring of fire'. Australia is devoid of active volcanoes, although in the eastern zone of the continent, from north to south, there is abundant, and geologically young, volcanic rock, and there are quite fresh-looking volcanic cones in parts of Victoria and South Australia. Nearby island nations — New Zealand, Papua–New Guinea, Indonesia and the Philippines — all have active volcanoes, and evidence of their activity reaches Australia from time to time in the form of vivid sunsets resulting from fine volcanic ash in the upper atmosphere, and tsunami washing the northern shores of the continent. The explosion of Krakatau (Indonesia) in 1883 was actually heard over all of northern Australia, and as far south as Alice Springs!

There is a variety of volcanic hazard phenomena (table 4.3), and in addition, interactions with other earth-surface processes give rise to secondary hazards. Lava flows of hazard significance average about sixty per century (Booth, 1979), but since they generally flow quite slowly along pre-determined courses governed by topography, they rarely pose a threat to life, although damage to property may be total. Lava flows frequently result in the sterilisation of large areas of previously productive landscape and it may be many years before weathering reduces the lava to fertile soil. On the other hand, lava flows may increase the area of land: the island of Hawaii is frequently enlarged by lava flows extending into the sea, and in 1973 about two square kilometres were added to the Icelandic island of Heimaey by a lava flow. The latter flow threatened to block the entrance to the only harbour of the island, an event which was prevented

Table 4.3 Principal volcanic hazards

Primary volcanic hazards	Secondary volcanic hazards
Premonitory earthquakes	Lahars
Flows of volcanic products	Secondary debris flows
lava flows	
pyroclastic flows, surges	Ground collapse and earthquake
debris flows	
debris avalanches	Flood (jökulhlaup)
toxic gas flows	
	Tsunami
Airborne volcanic ash	
	Acid rain
Fallout of volcanic products	
tephra	Atmospheric shock waves
ballistic projectiles	
	Post-eruption famine and disease
Explosions	
lateral blasts	
phreatic explosions	

Figure 4.4 The eruption of Vulcan. This remarkable sequence of photographs was taken by a crewman on board the *Golden Bear*, and records at close range the initial stages of the eruption of Vulcan, near Rabaul, in 1937 — at Long. 152.5°E, Lat. 4°S an eruption very close to Australia. The photograph at top left shows a small, dark, spire-like feature which is the beginning of ash ejection, and photographs 4 and 5 (left middle and lower centre) illustrate a pyroclastic flow surging from the vent, and spreading out over the water surface. As photograph

by spraying millions of litres of sea water on the advancing lava front so that the cooled and solidified outer crust of the lava would act as a containment wall (Williams and Moore, 1973).

Pyroclastic (the term is used for any rock material fragmented by volcanic action) flows and surges consist of fine particles of volcanic ash (material <4 mm in dia.) suspended in a highly turbulent gas flow. The gas may be any combination of air, volcanic gases, or water vapour, and while cold flows are known, they are usually hot, and frequently incandescent, giving rise to the common name of nuée ardente ('burning cloud'). The mobility of pyroclastic flows permits them to travel at velocities in excess of 150 km/h on steep upper slopes of volcanoes and they are usually hot enough to kill instantly by cadaveric spasm or heat asphyxiation (Baxter, 1990). They are capable of affecting limited areas only, but can cause high death tolls in densely populated regions. Pyroclastic surges from eruptions at Vulcan (figure 4.4) and Mt. Lamington in Papua–New Guinea killed over 3000 people (Taylor, 1958; Johnson and Threlfall, 1985), and from Mt Pelée in Martinique (Fisher et al., 1980) about ten times that number.

Lava debris, pumice fragments, ash, and scoria may avalanche, or even flow as a kind of slurry when eruptive processes repeatedly build up deposits with slope angles exceeding the angle of repose of the material.

Volcanic gases may vary somewhat in composition, but they always contain some constituents which may prove deadly to humans or animals if encountered in sufficient concentrations. Water vapour makes up the bulk of volcanic gases, usually about 70 per cent, but other constituents include carbon dioxide, carbon monoxide, hydrogen chloride, sulphur trioxide, hydrogen fluoride, as well as small amounts of methane, ammonia, hydrogen thiocynate and other rare gases. By virtue of being heavier than air they tend to move along topographic depressions and collect in low-lying areas. In 1986 there was a huge outburst of carbon dioxide gas from Lake Nyos, a crater lake in Cameroon, central Africa. In the immediate area of the lake many people were reported to have experienced a warm sensation, smelled an odour of rotten eggs or gunpowder, and rapidly lost consciousness. The gas, being heavier than air, flowed down

8 (lower right) was being taken, the captain became fearful for the safety of his crew, and ordered evacuation of the ship. However, as evacuation commenced, tephra fall over the ship reduced visibility to near zero. The crew crawled ashore, feeling their way, and, by following the line of the boards on the wharf, were able to locate a shed in which they took refuge. Photograph 7 (lower left) was taken later, and is not part of the eruption sequence.

river valleys, causing the death by asphyxiation of at least 1700 people, and thousands of domestic animals (Sigurdsson, 1987; Othman-Chande, 1987). There also may be secondary effects, as in the 1783 eruption in Iceland, in which the lava was accompanied by sulphurous gases which led to the deaths of large numbers of livestock and severely damaged crops. This in turn reduced the population of Iceland by some 20 per cent in the immediately ensuing years.

Volcanic ash in the atmosphere presents a hazard to jet aircraft: over 60 large aircraft have been damaged, some at distances of hundreds of kilometres from the eruption, and there have been some near-disastrous incidents (Casadevall, 1991). Pyroclastic fallouts of hazard status average about sixty per century (Booth, 1979) and are usually serious in their effects on life and property, especially where urban areas are near the source vent or lie along the axis of maximum dispersal. Rapid accumulation of ash during pyroclastic eruptions impairs the functions of both human and machine, and seriously hampers communications and evacuation (Peterson, 1990; Cook et al., 1981). Light falls deposit a thin layer of ash over the landscape surrounding the volcano and temporarily damage vegetation, but heavy falls may choke agricultural lands. Permanent damage to or death of plants occurs under moderate fallout, or where high concentrations of fluorine or sulphur dioxide gases are present in the ash. Ash may damage the teeth and clog the digestive systems of grazing animals, and high concentrations of poisonous elements in the ash may render grass lethal to livestock. The weight of ash may cause the collapse of roofs and incandescent ballistic blocks may cause fires when they crash through roofs or windows. The hazard of ballistic fragments is also of considerable significance to people without adequate head covering when they are attempting to evacuate.

Explosions and blast effects are minor but nonetheless important hazards which may result in destruction of forests, damage to housing, and even impairment of the hearing of people in the affected area. The writer saw large trees, up to a metre in diameter, spread over the landscape like fur on a cat's back by the blast from the Mount St Helens eruption in 1980. The level of danger of a volcanic explosion is partly determined from the point at which the blast issues. In the case of a crater eruption, most energy is directed upward and while tephra (collective term for clastic volcanic ejecta of any size) falls may be affected by wind direction the heavier pyroclastic fragments often fall back into the crater. Flank eruptions on the other hand result in the directing of the blast laterally and may occur if the vent of the main crater is plugged by solidified lava.

Lahars, or volcanic mud flows, are mass movement phenomena which consist of a mixture of water and volcanic ash: the high water content may be the result of heavy rainfall, of rapidly melting snow or glacial ice on which the ash falls, or even the bursting of a crater lake. Lahars may occur independently of volcanic eruptions but frequently follow them. Lahars are particularly important hazards on many volcanoes and are as destructive as pyroclastic flows over limited areas.

Structural collapse of a large volcano may occur as a result of emptying of the magma reservoir beneath, to form a caldera, and may also initiate landslip events. Caldera collapse alone does not usually pose a significant hazard, since a caldera forms near the summit of a volcano and subsides piston-wise in response to evacuation of the magma chamber. Landslips, on the other hand, often include large areas of adjacent lowlands. Tsunami are also associated with violently explosive volcanoes or caldera collapse in marine areas.

Eruptions that take place beneath ice sheets may cause meltwater floods (jökulhlaups): the increased water volume eventually raises the less dense ice hydrostatically and allows the sudden release of meltwater from beneath. Destructive floods and mudflows can be caused by sudden melting of ice and snow on high volcanoes.

The management of volcanic hazards

There are several entities involved in a volcanic crisis. The first is the source of the crisis itself, the volcano; next are those who study it, the teams of scientists; then those who decide what action to take to reduce the danger, the administrative authorities; those who divulge the information, the media; and finally, those who are exposed to the risk, the population who live near by. The behaviour of each of these parties and the relationships between them will determine the extent of the risk as well as the possibility of minimising it (Peterson, 1990).

Effective volcanic risk management requires that prediction should cover time intervals comparable with the time scale of human and social responses to them. The two principal responses are (1) appropriate land-use zonation, including location of settlements (time scale = decades) and (2) emergency evacuation of danger zones (time scale = days or hours). The first type of response may be achieved through zoning and risk assessment based on precise mapping of volcanic phenomena and products and dating of previous events (Booth, 1979; Tilling, 1989). The second demands the close co-operation of scientists and civil defence in order to define the levels of risk and corresponding social responses, to define the time scale and required accuracy of useful predictions, and to devise appropriate monitoring systems (Fournier d'Albe, 1980). Monitoring vital signals, especially seismicity, ground surface deformation, volcanic gas emissions, and eruptive phenomena may be desirable at potentially active volcanoes (Swanson and Kierle, 1988). Although damage resulting from volcano-seismic activity prior to an eruption is rare, the premonitory earthquakes themselves can serve as useful predictors of impending eruptive activity. Magma may move underground before erupting, and that movement is detectable. An accurate forecast of the 1984 eruption of Mauna Loa (Hawaii) was based on precursory seismicity and observation of slight swelling of the summit associated with magma movement.

Tsunami

Tsunami, or seismic sea waves, sometimes incorrectly termed 'tidal waves', result from sudden movements of the crust of the earth, as from earthquakes, large landslides into the sea, or volcanic eruptions. If a large area, maybe several tens or even hundreds of square kilometres, of the crust of the earth below the sea surface is suddenly raised or lowered, the displacement will cause a sudden rise or fall in the sea surface level above it. In the case of a rise, gravity causes the suddenly elevated water to return to the equilibrium surface level; if a depression is produced, the surrounding water will flow into it. In either case the result is the production of waves with extremely long wavelengths (100 to 200 km) and long periods (10–20 minutes), but very low amplitude. In addition to the magnitude of the causal event, maximum tsunami amplitude is related to depth of water above earthquake epicentre (Murty, 1977) — earthquakes below shallow water will not produce large tsunami. Tsunami may be compared to the waves generated by dropping a rock in a pond — waves (ripples) move out from the source in all directions. In general, the amplitude of tsunami waves decreases, but the number increases, with distance from the source region. Tsunami may be refracted, diffracted, or reflected by islands, seamounts, submarine ridges or shores. Tsunami velocity (V) is dependent on ocean depth (D):

$$V = \sqrt{gD}$$

where g = gravitational constant $\approx$ 9.81

For example, in the case of an ocean reach of depth 4000 m,

$$V = \sqrt{(9.81 \times 4000)} = 198 \text{ m s}^{-1} \ (= 713 \text{ km/h})$$

A tsunami is not likely to be detected by a shipboard observer in the open ocean, since its height of a metre or less is distributed over a

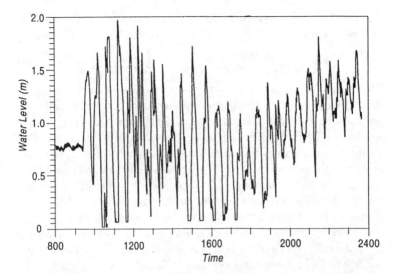

Figure 4.5 Hawaiian tide gauge record of the passage of the tsunami of 23 May 1960, which originated at Chile just after 2100 hrs the previous evening (Chile time).
Tide gauge trace courtesy CERC, US Army, Corps of Engineers.

wavelength of many kilometres, and is thus not easily seen or felt when superimposed on the other distortions of the ocean surface. It will be detected on a tide gauge record, however, as a quasi-periodic oscillation superimposed on the normal tide (see figure 4.5). The tsunami wave height, and subsequent runup, will be modified by interaction with local sea-floor topography and embayment configuration as it approaches land. When the tsunami reaches a coast, its energy is compressed into a smaller water volume as the ocean depth decreases. The rapid and sudden increase in energy density causes the wave height to increase rapidly. The resulting surge has the capacity to destroy most man-made structures in the runup zone, and may even carry quite large vessels well up on the land, leaving them stranded when the water recedes: the steamer *Berouw* was carried 2½ km inland by the 1883 Krakatau tsunami, and stranded 24 m above sea level. The same event killed 36 000 people, and destroyed over 5000 boats on the shores of Sumatra and Java, more than 50 km from the volcano. Most tsunami occur in the Pacific Ocean, which is ringed by crustal faults and volcanic activity, but they have also appeared in the Caribbean Sea, which is bounded by an active island arc system, and in the Mediterranean. Hawaii, located at a central point in the Pacific, records about one tsunami per year, on average, although few cause damage. The 1960 event (see figure 4.5) produced maximum runup of 10.7 m at Hilo, and resulted in damages of over US$100 m (1993 values).

The management of the tsunami hazard

The obvious precaution of restricting development on Pacific shores at risk is generally uneconomic, given the infrequency of tsunami, although the zone of maximum risk at Hilo, Hawaii, has been zoned as public open space since 1960. As a tsunami takes about twenty hours to cross the Pacific, a monitoring system is used to alert emergency services personnel in vulnerable areas so that coastal evacuations can be organised. Information from seismic stations, and a series of sea-level indicators in the Pacific are used in conjunction with computer models of wave behaviour to predict arrival times accurately. An educational brochure explaining evacuation, originating from New Zealand, bears the title *TSUNAMI — If you see one it will be too late!* People tend to become blasé, however, and periodic re-education and drill exercises are in order. Although the population of Hilo had ample warning to evacuate in 1960, 61 people were killed and 282 injured. Many became bored with the apparent inaction while waiting in the hills, and returned, while there were those who stayed 'to watch the fun'.

In areas most at risk, microzonation of the potential runup zones, based on historical data and mathematical modelling of tsunami behaviour, with subsequent rationalisation of land use, appears desirable. Based on microzonation maps, redistribution of population, structures and services may be appropriate, and the maps may be used to plan the location of shelters and evacuation routes. Within runup zones, elevation of buildings on strong piles, protective barriers against wave force and wave-borne debris, and forested buffer strips (to dissipate tsunami energy) have been used. On some tsunami-prone coasts of Japan, seawalls several metres high, fitted with gates to allow waterfront access under normal conditions, are in place (Iida and Iwasaki, 1983).

The Australian continent is relatively unaffected by tsunami. Tsunami originating in Chile, Japan, and Alaska have produced water level perturbations of up to 1.2 m on the eastern Australian coast, while the highest tsunami runup of 6.0 m ever recorded on the Australian coast (at Cape Leveque, 16.24°S, 122.56°E) was of a 1977 tsunami originating from a shallow 8.0 magnitude earthquake just south of Sumba Island, Indonesia. The same event was observed widely in northwestern Australia, and also created runup of up to 15 m in the Lesser Sunda Islands, close to the earthquake epicentre.

Mass Movement Phenomena

Gravitationally induced downslope movement of earth surface materials, or of snow or ice, are of fundamental importance in shaping the surface of the earth around us. Some, such as soil or talus creep, or glacial movement, are too slow to be observable without instruments: we are concerned here with the hazards of landslides or avalanches.

Landslides

A landslide is the rapid movement of a mass of earth material due to slip surface failure (along which the slide occurs), when the *shearing stresses* exceed the *shear strength* (Crozier, 1986). Causal factors are summarised in table 4.4. There is a continuum between landslides and mudflows with behaviour of material largely determined by water content, and particle size and sorting. Management measures usually adopted fall into categories of *avoidance, non-conflicting use,* or *engineering control.* Avoidance is best practised where the degree of risk is great, and the area may be left undeveloped, or utilised for forestry, some types of agriculture, or as parkland. Some non-conflicting uses may be allowable where strata are not overloaded or undercut, or liable to saturation of substrata. Engineering controls may be employed where economic pressures, and land limitations, force use of moderately unstable areas (as in parts of Japan, and Hong Kong, for example).

Australia is remote from tectonic plate margins and is characterised by old, hard rocks and an arid climate. Landslides are generally newsworthy because of their rarity rather than their destructiveness, although there are certain strata prone to mass-movement, especially along the moister eastern seaboard where geomorphic processes are more active (Ingles, 1974; Dunkerley, 1976; Joyce, 1979; Donaldson, 1980; Hollingsworth, 1982). Roads and railways have been damaged, and housing affected, by slope movement, especially in the Illawarra district of New South Wales.

Avalanches

Snow avalanche is a type of slope failure that can occur when snow is deposited on slopes steeper than about 20 to 30 degrees. Avalanche-prone areas can be delineated with some accuracy, since under normal circumstances avalanches tend to run down the same paths year after year, although exceptional weather conditions can produce avalanches that overrun normal path boundaries or create new paths. Unlike other forms of slope failure, snow avalanches can build and be triggered many times in a given winter season.

Table 4.4   Landslides — causal factors

Conditions favouring landslides

Lithologic — weak formations such as unconsolidated materials or poorly cemented sediments.

Stratigraphic — massive beds overlying weaker beds or permeable beds, lenses or wedges of sand, or other porous material.

Structural — dipping planes of weakness, such as fault planes, joint planes, cleavage planes; fractured, crushed or slicken-sided rock material.

Topographic — over-steepened slopes from glacial, fluvial or wave erosion, block faulting from previous landslides, or from human modification of slope (benching for homesites, road cuttings); weakening of slopes by removal or destruction of retaining vegetation by deforestation, overgrazing, or cultivation.

Landslide initiating factors

Earth vibrations — from earthquakes, collapse of caverns or mine workings, explosions, heavy machinery.

Removal of support — outflow, or decrease in volume, of subjacent layer of plastic clay or quicksand; burning out of coal or lignite bed.

Overloading — rockfall or slide onto upper part of slope; saturation of regolith; artificial loads such as houses, reservoirs, above-ground swimming pools, earthfill.

Reduction of friction — lubrication of slip plane, or softening of weak or unconsolidated rock, by oil seep or chemical alteration, or most commonly, by water (poor drainage, septic tanks, leaky swimming pools, water mains or reservoirs, overwatering of gardens).

Reduction of cohesion through desiccation, or disturbance of certain rheologic clays.

Prying or wedging action — freezing water in fissures; tree roots in fissures; hydrostatic water pressure in joints; salt crystallisation in joints; swelling of colloids; temperature expansion.

There are four basic prerequisites for an avalanche to develop: (1) the accumulation of a critical mass of snow; (2) structural changes within the snow which reduce the stability of the mass; (3) an adequate slope angle to permit a gravity flow, and (4) a triggering mechanism. The amount of snow necessary to produce a critical mass can not be stated in absolute terms because snow is of variable density, snowflakes attain different shapes and mass depending on the conditions in the atmosphere under which they form, and undergo further change as they strike the ground and are buried under additional layers of snow. In addition, the building of a snow pack resembles the process of sedimentation since the deposition of layers of different type of snow, and the subsequent weather conditions affecting those layers, result in a stratified body which may contain several planes of weakness.

Snow avalanches are of little consequence in Australia, but pose a mounting hazard as development and recreation increase in some mountain areas, especially the European Alps, North American Rockies, and the south island of New Zealand. Hazard mitigation requires measures ranging from appropriate land-use management and effective building codes in avalanche-prone areas to the timely issuance of emergency warnings and programmes of public education (Mears, 1992).

CONCLUSION

Throughout civilisation earthquake damages have been locally catastrophic, but have weakly influenced the course of history (Bilham, 1991). However, the growth of mega-cities has resulted in large social and economic targets for future earthquakes, and the construction of unreinforced structures within many of these cities continues. Bilham (1988) pointed out that there are at least forty cities with populations exceeding 2 million located within 200 kilometres of a historically damaging earthquake. The population within this handful of cities is anticipated to reach 600 million by the year 2035. Earthquake-resistant building technology is available, but the lessons from Armenia and other places show that economic or political pressures for inexpensive construction may prevail over the political will to enforce stringent building regulations, to the detriment of hazard mitigation. There are also huge numbers of people living in structures which have been built without reference to any kind of building code, and which, especially where adobe is the preferred construction material, pose a large, but unspecifiable, catastrophe potential. Growth in world population, therefore, and the observation that much of this population will reside in improperly designed urban dwellings within or near seismic zones, leads to the inescapable conclusion that the loss of life and property from earthquakes in the next century is likely to exceed that in all recorded history.

Significant advances have been made in techniques for volcano monitoring, hazard assessment, and eruption forecasting. As with the earthquake hazard, the major problem in mitigation of volcanic risk on the global scale is that most of the dangerous volcanoes of the world are in densely populated areas of countries that lack the economic and scientific resources or political will to study and monitor them adequately. There are great advances to be made from the application of existing technology to the poorly understood, little studied volcanoes in such areas — probably greater than might be gained from refinement of present technologies or development of new methods for volcano monitoring, hazard assessment, and eruption forecasting. There is also the question of public education with regard to a hazard which is particularly prone to interpretation through filters of fear, superstition, or myth by laypeople. Scientists customarily place priority on observing, monitoring, and interpreting a restless volcano; by default, a lower priority is generally placed on explaining their findings to members of the public. Furthermore, they may become understandably impatient when responding to naïve or misconceived questions about the volcano from civil authorities, the news media, or citizens, but such an attitude is almost certain to produce controversy and misunderstanding. On the other hand, when officials and reporters fail to become informed about easily acquired basic facts, or when they demand definite answers when such answers are impossible to provide, they add to the potential for confusion and poor decision making.

■

5

Floods — Water, water, everywhere

From the human point of view, a flood can be defined simply as water where it is not wanted. Flooding is a normal part of river behaviour. The technical definition of a flood is arbitrary, but is usually associated with some stage or height of water above a given datum such as the banks of the normal river channel. The channel is not adapted to carry all the water delivered to it all of the time: excess water is carried on the flood-plain of the river. Geomorphologically, the floodplain is a part of the river bottom, which is a fact of great importance when the impact of flooding upon society is considered.

Floods can be caused by a number of factors, singly or jointly. Common natural factors are: intense storm precipitation, high antecedent basin soil moisture, rainfall over areas covered with snow, rapid snow-melt, occurrence of medium to major storms in quick succession, and failure of ice dams resulting in a very rapid release of large quantities of water. Floods may also be induced by earthquake or landslide. In addition, some floods may be wholly or partly caused by people, in the case of those caused by dam or levee failure, by the modification of river catchments, or by land subsidence induced by over-pumping of aquifers or withdrawal of oil from shallow subterranean sources.

The history of all nations includes stories of major flood disasters, and Australia, despite its youthfulness as a nation, is no exception (figure 5.1). There are some 200 000 urban properties prone to flooding by the

Figure 5.1 Urban flash flooding is a significant hazard in parts of all of the major Australian urban centres. Photograph shows flash flooding in Elizabeth Street, Melbourne. The VW in the picture is being carried along on the flood-waters — it disappeared a few seconds after the photograph was taken.
Photograph courtesy Bureau of Meteorology, Australia.

100-year Recurrence Interval event, and, on average, flooding costs the nation $300±50 million per year, with $50 million being spent on flood relief and about $18 million on mitigation works and flood studies (AWRC, 1992). In the period 1950–52, flooding in much of the Murray-Darling and Hawkesbury-Nepean systems destroyed over a third of the wheat crop and caused other damages estimated at over £22 million (about $370 million 1993 dollars). In 1973–74, heavy rain fell over most of the continent, Lake Eyre filled for the first time in history, and in the Brisbane-Ipswich area alone, flooding caused almost $200 million in damages. In Australia, floods cause few fatalities, but this is not true, unfortunately, for countries such as India (particularly the northern states) and Bangladesh, where over 300 million people live in areas that may be affected by floods in any year, and, on average, over 3 000 people and 100 000 head of cattle perish each year in swollen rivers (as distinct from storm surge, or sea flooding: for which see chapter 3). Floods are, however, not necessarily solely destructive. The Egyptians living along the Nile River considered floods a life giving flow which spread fertile alluvium and moisture in which crops could flourish.

If it is true that the floodplain belongs to the river and that floods are bound to happen, then why are floodplains occupied at all? The reasons include the fact that in the early days of settlement rivers provided water supply and transportation, and sometimes power, for growing settlements. Floodplain land provided good farmland as well as easy routes for transportation. In some regions the only flat land available for settlement and agriculture lay in the valley bottoms. Despite repeated devastation, people are often reluctant to move because they perceive that the advantages of location, ranging from economic to aesthetic, outweigh the disadvantages of flooding. At the same time public policies, from subsidised rehabilitation of damaged property, to reservoir and levee construction at the general taxpayer's expense, have encouraged exposure to floods in many areas.

FLOOD FLOWS AND THEIR MEASUREMENT

The measurement of actual flood flows is usually accomplished by means of a *discharge hydrograph* (figure 5.2) which expresses the outcome of the relationships that occur between runoff and the other components of the water balance of a basin, together with their adjustments to the physical characteristics of the catchment. The shape and dimensions of the hydrograph for a particular catchment are controlled by a variety of factors, many of which are interrelated (figure 5.3). In the case of rain-induced floods the path of the storm in relation to the alignment of the basin, the spatial extent of the storm, and its rate of movement are all important, as well as the intensity and total amount of rain. Once the rain has reached the surface of the basin, the rate and amount of runoff will be influenced by factors such as the current evaporation and infiltration rates, soil moisture status, and type of land use. Movement of water into channels of increasingly higher order, and the passage of the flood wave, is governed by a series of factors, from the configuration of the basin to the hydrologic and biological characteristics of the channel system, which control not only the form of the hydrograph but also the time interval between the rain and the flood. *Basin lag* and *time of concentration* are indices of the time response characteristics of a catchment, the first being the time between the centre of mass of rain and centre of mass of runoff, and the second the time taken for water to reach the gauging station from the most distant point in the catchment. The *unit hydrograph* is frequently used for prediction of flood flows. It assumes that the runoff from uniform effective rainfalls of the same duration, produced by isolated storms on the same basin, causes flood peaks on the hydrograph of equal length in time, with the ordinates (Y-axis) of the flood hydrographs being proportional to the total volume of direct runoff from falls of equal duration, but

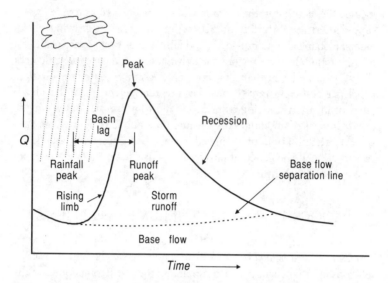

Figure 5.2 Hydrograph curve. When the quantity of runoff from a catchment is measured continuously, a graph of flow as a function of time, known as a hydrograph, is obtained. During dry periods, the base flow decreases exponentially. The base flow continues until rain falls. The typical hydrograph exhibits a rapid increase at some time following rain, created by the convergence of channel precipitation, plus overland flow (surface runoff). The flood peak for a given basin occurs at some defined time following the rainfall peak — this time period is known as the basin lag. The slower recession of the hydrograph reflects the fact that the delivery of water to the stream channel from interflow pathways, regolith storage, and groundwater storage is delayed by comparison with the rapid arrival of overland flow.

different intensity. To derive a unit hydrograph for a particular basin, records of rainfall and discharge are examined for a typical isolated storm with uniform rainfall. For such a storm, the base flow is separated from the remainder of the hydrograph so that the volume of storm runoff can be determined. This is then expressed as the hydrograph from a unit (usually one millimetre) of rainfall. For example, given a hydrograph resulting from uniform rainfall of 2.3 mm over an entire catchment, the unit hydrograph would be arrived at by dividing each ordinate of the storm runoff hydrograph by 2.3.

Effective flood management demands an understanding of the magnitude/frequency relationships of flow for a particular catchment, and how these are likely to be modified, for better or worse, by human intervention. Perfect knowledge never exists, and a variety of methods have been used to try and predict the response of catchments to flood events. The

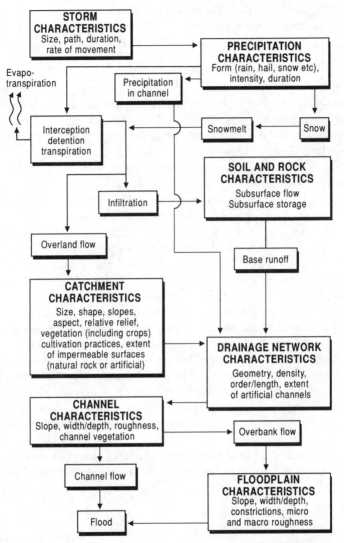

Figure 5.3 A summary of the factors involved in producing runoff, stream channel flow, and flooding. The drainage basin, a part of the surface of the lithosphere that is occupied by a drainage system, or contributes surface water to that system, is the fundamental unit of analysis for hydrology and fluvial geomorphology, and also for the management of flood behaviour. In any natural drainage system, the runoff channels are organised into a coherent network, and the system is finely adjusted so that there are mutual, interdependent relationships between the input and output of water; weathering and sediment production; sediment transport and sediment throughput; vegetation type, and slopes.

methods fall into three broad (and overlapping) areas: (1) statistical — based either on hydrograph records or on rainfall records and catchment characteristics; (2) empirical, and (3) digital simulation models.

Statistical methods

Statistical methods usually involve fitting a frequency distribution to a sample of flood observations from a given stream. Ideally, the largest flood from each year of record is used. The frequency distribution is then extrapolated if necessary to the probability level of interest. Probabilities for flood flows, and indeed most hazard events, are usually expressed in terms of Recurrence Interval (or Return Period): the *average* period of N years between events of specified magnitude is termed the Recurrence Interval. The largest flood flow in each of thirty years, say, may be used to construct a Recurrence Interval plot on special *Gumbel* or *Log-normal* plotting paper, on which the relevant frequency distribution forms a straight line that may readily be extrapolated to Recurrence Intervals of, say, 50 or 100 years. With respect to floods, length of hydrograph record is critical, since extrapolations become increasingly less reliable the greater the extrapolated period in comparison with the period of record.

There are also statistical methods based on rainfall records and catchment characteristics, which are often the only methods available for use with small ungauged catchments or in new areas (e.g., new residential subdivisions), where stream discharge information may not be available. If a reasonably long sequence of rainfall records is available from the catchment in question, or from stations nearby, it may provide statistically valid estimates of *input* to the flood event. All such methods include a basin area factor and some index of rainfall intensity, in addition to measures of landform, land use, and soil type characteristics (which may be summarised in a runoff coefficient). The simplest of these involves the use of the so-called Rational Formula for the estimation of the discharge (Q) for a given Recurrence Interval (RI):

$$Q_{RI} = 0.2778 \, CI_{RI}A$$

where

C = the runoff co-efficient ($0 \le C \le 1$),
I = rainfall intensity, for the recurrence interval (RI) specified, and
A = area in square kilometres.

In this case, I has units of millimetres per hour, and Q will have units of cumecs (cubic metres per second).

Factors that have commonly been used in more complex models include: a topography factor (slope angle, slope length, relief); a rainfall frequency factor; main channel slope and/or main channel length; a temperature factor; and drainage density. There is an obvious relationship between the size of the catchment, the amount of rain it may accept, and the peak discharge that results, but basin shape is also of importance in influencing peak flow and other hydrograph characteristics: for two basins of the same area, the flood potential would be considered greatest for the one most closely approximating a circle in shape.

Human modification of catchments is critical in flood behaviour. A drainage basin may be considered as a system, with an *input* of a precipitation event which produces an *output* of a flood event. The *system variables,* such as land use, are responsible for the nature of the flood which is produced by a given precipitation event. Land use changes, from forest to cultivated land or grassland, or from either of these to urban, will significantly affect the behaviour of floods. Urbanisation usually produces an increase in the magnitude of overbank flows (or a reduction in the Recurrence Interval associated with overbank flows of a given stage), and reduction of the lag times. Modifications to system variables may be summarised as: the extension of impermeable surfaces; the construction of artificial channels, and the increase of drainage density.

Empirical methods

Empirical methods usually involve some kind of estimate for probable maximum precipitation which is then used in conjunction with either the Rational Formula as above, or in conjunction with the discharge records from catchments of similar characteristics. The process of deriving an estimate of probable maximum precipitation (PMP) normally involves: the identification of maximum recorded rainfall(s) for the specific catchment and/or other comparable areas; the transposition of those storms, where this is appropriate, to the catchment under consideration, and the possible upward adjustment of the transposed rainfall values. The principal upper limits on precipitation which must be quantified are: the maximum humidity concentration of the air; the maximum rate at which humid air can move into a specified area, and the proportion of the inflowing water vapour which can be precipitated. Where good streamflow records are available from analogous areas, a discharge/area graph may be used to extrapolate the probable N-year flood — in the limit, the maximum observed floods from large catchments all over the world may be fitted by an envelope curve, or by a line-of-best-fit (figure 5.4).

Digital simulation models

Digital simulation models are designed to reproduce mathematically the components of the catchment system. Digital models carry out two principal operations: converting precipitation events into runoff, and transforming volumes of runoff into discharge hydrographs. Parameters in the programs may be readily altered to represent any particular set of circumstances. Time scales can be compressed, and the simulated behaviour of a basin over a period of years can be reproduced in a few minutes. We often use models in science to abstract and study processes. In designing a model, we must first resolve the conflict between the span of the model (how much of the real world it will represent), its resolution (minimum process differences that can be detected), and its size and complexity (number of variables times the number of interactions). As any of the trio becomes more demanding or limiting, one or both of the other two must be proportionately compromised. Space, time, and process scales are inextricably linked, and models that have inconsistent scales are unlikely to be either efficient or accurate. The effort to calculate the details is wasted in studying general phenomena, but general models are too coarse to detect detailed processes of interest. Global-scale models, for example, are of limited value for examining regional phenomena. Much work has been put into the development of operational mathematical models for river-flow forecasting: the most sophisticated employ some kind of mathematical formulation of the conditions and linkages illustrated in figure 5.3. The advantage of models based on a physically sound description of catchment attributes and hydrological processes is that their parameters have direct physical interpretation, and values may be established by field or laboratory investigation (Kachroo, 1992).

Digital simulation models are used in actual flood forecasting schemes, where stream-flow forecasts are required in order to issue flood warnings, and to permit the timely evacuation of people threatened by rising water. Discharge forecasts are obtained in real time, by using the model to transform input data (actual or forecast rainfall, or flow measurements at upstream points) into discharge as a function of time. Digital models are also used at the design stage of hydrological or flood control works, and to aid efficient operation of reservoirs for hydro-power or irrigation purposes.

FLOOD IMPACTS AND THEIR ASSESSMENT

Flooding does not simply present a single perceived risk but a variety of risks, including: the risk of a flood *per se*, with associated direct damages

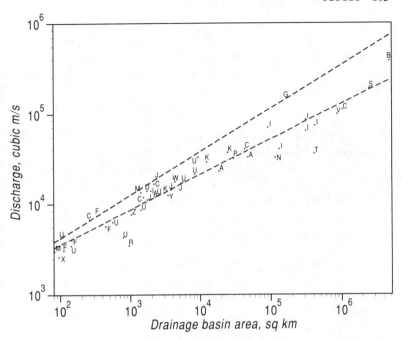

Figure 5.4 Maximum world floods in relation to area of drainage basin. Large catchments usually have less steep average slopes than do smaller catchments, and it would be highly improbable for rainfall from a single storm to affect the entire area of a very large catchment, especially where the catchment includes different climatic zones, such as the Murray-Darling system in Australia. Hence, the exponents in the formulae for the curves shown have a value of much less than unity (in other words, the discharge is not simply proportional to area) — the line of best fit through the data points (lower curve) is described by $Q = 604 A^{0.388}$ and the envelope by $Q = 470 A^{0.479}$.

A = Australia	M = Mexico	W = Philippines
B = Brazil	N = Vietnam	X = Cuba
C = China	P = Pakistan	Y = Malaysia
F = France	R = Italy	Z = New Zealand
G = Germany	S = Siberia	
I = India	T = Thailand	Data from Rodier and
J = Japan	U = USA	Roche (1984) and Baker
K = Korea	V = Venezuela	and Costa (1987).

(including loss of life); the risks to the environment and to amenities associated with any scheme to alleviate flooding (including the risk of scheme failure); the risk of panic in the event of a flood, and of looting in a flood; and even the risk to property prices of broader public knowledge that some areas are prone to flooding! If the risks associated with floods are varied, so too are floods themselves. While flooding results from intense local rainfall, from river surcharge, overloaded storm drains, the failure of natural or man-made dams, or from the sea, categorisation in terms of source is not necessarily related to similarity of effect: that is, to the risks and consequences of flooding.

The direct impacts of floods are related to (1) the depth of inundation; (2) the velocity of the flood flow; (3) the duration of inundation, and (4) the sediment load, floating missiles, or other substances carried by the floodwaters, including pathogens and pollutants, or, in the case of sea-water flooding, salt. The hazard for structures in the floodplain is evaluated with hydrologic, hydraulic, engineering, and economic information, which is used to delineate areas subject to inundation and to estimate flood flows and the severity of floods of varying frequency. Data used to quantify the flood hazard include: (a) anticipated depth of flooding, based on historical or design floods, and/or known frequency of flooding to some specified elevation; (b) damage resulting from historical floods, or expected from design flood(s) of specified depth; and (c) existing levels of protection and expected annual value of flood damages. Expected annual flood damage is defined as the average damage that can be expected in a year, and is computed by summing the products of the damage caused by each depth of flooding and the probability of each of those depths occurring.

Techniques to assess direct damages from river flooding are well developed and various computer programs are available (Penning-Rowsell et al., 1987). One of these, ANUFLOOD, was developed in Australia and has been used for several years for a variety of damage-assessment problems (e.g., in the southwestern suburbs of Sydney: Smith et al., 1990). There are three inputs: a land-use database, a series of river stage-damage curves, and information on flood probability. The land-use component is based on an external field assessment of all individual buildings: the survey, divided into residential and commercial sectors, records grid references, ground and floor heights, number of storeys, construction material, size, and, for commercial or industrial properties, use. Each building is assigned to a specific value category based on the susceptibility of the contents to flood damage (there are, for example, five damage categories for commercial buildings). ANUFLOOD also allows for the form of the flood surface (it is not 'flat'). When land use, stage-damage curves and flood probabilities are known, the program can provide damage estimates for each sector, for any selected flood height or probability (Smith, 1990).

With regard to loss of life from flash flood or dam failure, Brown and Graham (1988) describe a model which begins with the cognizance of the threatening conditions, such as a very large rain storm (or earthquake, say) and continues through event detection and monitoring, decision making on the part of responsible officials, the warning process, the response of people to the warning, the evacuation, and the impact of the flood on people remaining in the floodplain. Brown and Graham concluded that the three factors which most directly determine loss of life were the size of the population at risk, the amount of warning time available before flooding and the depth/velocity relationships of the flood waters. Provided adequate warning time was available (greater than $1\frac{1}{2}$ hours), the analysis showed that the probable loss of life from such an event, if properly handled, was equal to 0.0002 of the population at risk.

In Australia, the rivers of the large inland drainage systems are long, and flanked by extensive floodplains. During major flood events, vast areas are inundated by shallow, slow-moving floodwaters: a flood warning period of several days to several weeks is usually available in which evacuation of people or property may take place, although stock losses may be serious. On the other hand, the short steep coastal rivers, with their narrow floodplains, are liable to deep and swiftly flowing floodwaters, and warning times are short: 1–2 days maximum, often only a few hours. Although there may be a lead time for preparation in the case of inland flooding, settlements may be cut off with the only access being by air. The whole community may be affected, with its infrastructure virtually inoperable, and even the emergency co-ordinator's office may be under water, as was the case in the Charleville (Qld) floods of 1990. Urban flash flooding is a significant hazard in parts of all of the major Australian urban centres. Urban catchments are generally small, with impervious and 'hydraulically smooth' surfaces, and often have a large proportion of steep slopes (Bureau of Meteorology, 1985; Riley et al., 1985; AWRC, 1992).

MANAGEMENT OF THE FLOOD HAZARD

Flooding is managed on three fronts: there are measures designed to change the behaviour of the flood itself, such as dams or levees; there are measures designed to minimise the risk of damage from the flood event, as by land-use zoning to prevent occupance of high hazard risk areas, or regulations for minimum floor levels; and there are response measures, aimed at inducing effective reaction by flood-liable communities, such as flood warning and evacuation plans.

Modification of the flood hazard

Hazard modification is concerned with attempting to change the characteristics of the natural event so that its impact on human activities is diminished or nullified. Modification of the flood hazard usually begins in the catchment. Small dams in headwater streams, contour banking or terracing, gully control, modifications to cropping or ploughing practices, bushfire control, revegetation or reafforestation are all techniques which slow down the overland flow from a given rainfall event, increasing the lag time and allowing for more complete infiltration of rainfall into the ground surface. There are also channel modifications such as river diversions, the creation of artificial levees or the enhancement of natural levees, the construction of large dams (deliberately left with surplus capacity to accept flood flow), channel modification such as channel straightening, desnagging etc., and the construction of flood buffers or pondages: all methods which may be used to modify the flow of water after overland flow has taken place and the water enters the stream channel.

Avoiding the flood hazard

Hazard avoidance is concerned with location of human activities in space (planning) or in time (programming) so that they are unaffected, or relatively unaffected, by the hazard. However, a floodplain is a desirable location for human activity from many points of view. Historically, the location of trade centres on rivers was associated with the ease of transport by water, and the limited number of crossing points available on a river stimulated the development of urban centres at or close to those foci of human activity. In addition, the floodplain soils are usually richer than those on the surrounding non-floodplain lands and the development of agriculture on floodplains encouraged human settlement. Complete abandonment of floodplain settlement is usually not economically viable: there is a trade-off between the gain in safety from moving off the floodplain and the short- and long-term costs incurred thereby. However, suitable scheduling of activities, especially in the case of a predictable flood regime, such as from snow-melt, may be entirely feasible: the annual flood of the Nile became the basis of the cycle of human activity on the Nile floodplain and delta. In addition, many floodplain lands can be, and often are, used for activities which may be curtailed during times of inundation with minimal economic loss or inconvenience, such as for sportsfields.

Prevention or minimisation of loss from flood

Loss prevention is based on acceptance that the natural event will occur, and that even with hazard modification techniques such as those above, it is wise to take measures to minimise damage to, or loss of property. Flood damage to structures depends on: the type, strength and elevation of the structure; the depth of the floodwaters; the force exerted by the floodwaters; the impacts of floating debris; the wetting effects of the floodwater; and the load of silt or other material carried and deposited by the floodwater. Techniques commonly employed for loss prevention and minimisation include:

- making lower levels of a structure impervious;
- use of waterproof shutters or hatches which may be fixed over window or door openings when flood is predicted;
- waterproofing of the building fabric, including seepage control (plastic wrap or other waterproof material may be permanently installed for covering sensitive equipment);
- use of water-resistant materials in new structures;
- use of readily removable fittings;
- use of non-return valves in sewer lines;
- elevation of the substrate, as with fill, or raising structures in place, as on stilts;
- constructing flood walls or levees around structures (sand bagging or other temporary water barriers may also be used).

There are also non-structural measures for loss prevention or minimisation, such as: development of flood forecast and warning systems with appropriate evacuation plans; regulation of the development of floodplain land by use of zoning ordinances, subdivision regulations and building codes; acquisition of title or easement to floodplain land by appropriate authority (and possible relocation of existing structures or contents, or both, out of a flood hazard area); placement of warning signs in the floodplain to discourage development; and the use of tax incentives to encourage wise use of floodplain land. Formulation of successful non-structural flood-control plans requires integration of information on the flood hazard, the engineering characteristics of the floodplain structures, and the environmental and social suitability. Clearly the flood hazard, engineering, and environmental information are important, but the social and political suitability of a proposed plan will ultimately dictate its acceptability (Ford, 1981).

Sharing of losses

Loss sharing involves the acceptance that the natural event will occur, that losses are inevitable, but that impacts on individuals may be minimised through sharing arrangements. Loss sharing along family or kinship networks is a common response to flood (and other hazard) damage in many societies, and in developed countries it is common for governments to assist people in distress from flood. There can be no argument against disaster relief from funds built up by voluntarily contributions from citizens, but the provision of government handouts to flood victims has frequently been criticised, on the grounds that the taxpayer generally is subsidising (or even 'rewarding') the unwise actions of a few. It is unfortunate that government handouts, and other forms of relief, such as funds from public appeals, may come to be regarded as a 'right' by people living on floodplains, which perpetuates the situation of unwise land use. It has been recommended that the availability of relief payments in Australia should be made conditional on instigation of appropriate floodplain management (AWRC, 1992).

The question of insurance on flood-prone lands is also a difficult one, as the risk is quite site-specific, and is not one which applies to citizens generally. It may be economically viable in the long term for a certain activity to locate on the floodplain (if the long term benefits of a floodplain location exceed the long term flood costs), but the risk of large flood losses in the short term may deter the development. Insurance, by spreading these short-term costs over a longer period and thus minimising risk, may allow the development to take place, and thus may encourage unwise floodplain encroachment. Realistically, flood insurance should cost at least as much as the expected annual damages plus administrative costs (Arnell, 1987), but even so, the cost of such insurance may not deter floodplain encroachment if the location offers additional non-quantifiable benefits such as fishing rights, river access, or prestige value. In Australia, flood insurance is not usually available. It may not be generally affordable if limited to the population in flood-prone areas, but would not be equitable if spread across the wider population (AWRC, 1992).

CONCLUSION

Flooding is one of the more manageable of natural hazards. Windstorm may strike anywhere, but floods are restricted to relatively small, definable areas; forest products cannot be had without fire-prone forests, but many of the benefits of the floodplain may be enjoyed while at the same time minimising the risk from flooding. However, flooding remains a major social and economic problem in virtually all countries. It was Hegel who first observed, nearly two centuries ago, that experience teaches that we do not learn from experience, and one is tempted to say that nothing has changed. Floodplains are by definition flood prone. Minor floods are relatively common in most river valleys, and provide warnings of bigger things to come. The causes of floods, and flood behaviour, have been well studied for many years. And yet many urban areas were not developed in a manner which took cognizance of potential flood risk, or of the effect of new developments on flood behaviour. In Australia alone, there are an estimated 80 000 properties subject to stormwater flooding and another 120 000 subject to mainstream flooding from the '1 in 100 year' flood event. In large catchments, floodplain development and management activities have generally been fragmented and poorly co-ordinated.

Areas at risk from flood can readily be determined, and techniques for assessing the vulnerability of existing or proposed structures, and of the probability of flooding, and consequently of risk, are well developed. There has been sufficient experience with mitigation measures for us to determine the likelihood of their effectiveness in most applications, and of the consequences of catastrophic failure when the design event is exceeded or maintenance disregarded. The lesson of ten billion dollars worth of damage in the catastrophic July 1993 Mississippi floods in the USA stands as evidence of the consequences of failure of mitigation works, as in this case most damage was due to overtopping or failure of levees by record or near-record events (a well-illustrated account appears in the January 1994 *National Geographic*). It would seem therefore, that the most effective way of limiting future flood damages is by the adoption, and strict enforcement, of appropriate planning controls. Floodplain lands are a valuable resource, and will continue to be so, especially for their agricultural potential. Effective management plans will recognise that the floodplain is, by nature, from time to time part of the river bottom, and will promote those land uses which can best take advantage of the fertile soils and proximity to water that the floodplain provides, while at the same time being prepared to pay the cost of these advantages, which is occasional inundation.

■

6

Drought —
...and not a drop
to drink

What is a drought? Tannehill (1947) observed that 'we may truthfully say that we scarcely know a drought when we see one.' Unlike other hazard phenomena, drought is not readily recognised by the physical symptoms alone: rather, it is usually defined in terms of outcomes. Especially in the poorer or less well-developed parts of the world, drought frequently brings famine in its wake. The oral and written histories of ancient civilisations provide ample evidence of recurrent drought and famine. A 'stele of famine', dated at ca. 3500 BC, and found in Egypt, tells of the hardships of a famine caused by a seven-year failure of the annual Nile floods: 'torn open are the chests of provisions but instead of contents there is air', and a dark chapter in Indian history records the drought-provoked famine of 917–918 AD when thousands of corpses floated down the Jhelum River, and 'human bones covered the land in all directions' (Cornell, 1976). More recently, in the 1980s, drought-related famine has been by far the major cause of death due to natural disaster, with well-documented events in Africa (principally in the Horn of Africa) responsible for millions of deaths. Tragically, the effects of famine have been compounded in these countries by civil strife, and by political ideologies which have forced sedentary lifestyles upon erstwhile nomadic peoples, prohibiting move-ment between pastures which buffered effects of drought in the past.

Drought can be conceptually distinguished from most other natural hazards in two ways. First, drought is a natural hazard that is pervasive in nature; it is a creeping phenomenon, the effects of which are slowly accumulated and may persist over long periods of time, in contrast to the sudden and short-lived physical impact of hazards such as floods, tropical cyclones, or earthquakes. As a result, its effects are slow to be comprehended, and by the time intervention is sought, it may be too late for relief organisations to respond effectively. Second, the effects of drought — at least agricultural drought — are often woven into the economic and social fabric of a region or a nation (Wilhite and Easterling, 1987). Although droughts can have major effects in developed countries, the drastic effects in terms of human starvation are avoided and there are often efforts to improve land which has become badly eroded. On the other hand, thousands die and millions of people have been, and continue to be, affected by famine resulting from droughts in poorer countries, such as those in much of Africa in the 1980s and 1990s, and the difficulties of improving the badly degraded soils are very great. Drought may have other far-reaching consequences, in terms of population reduction and alterations of social structure, population shifts, economic hardship, and significant environmental perturbations (Bernard, 1985).

THE IDENTIFICATION OF DROUGHT

Drought is essentially a temporary shortfall of water supply below demand caused by the behaviour of natural atmospheric and hydrologic processes, and which has significant social and economic repercussions. It is not restricted to countries normally perceived as being arid or semi-arid: much of Europe was affected in the period 1988–92, with the drought in the UK being the most severe in the twentieth century (Marsh and Monkhouse, 1993). Drought should be distinguished from man-made *water shortages* created by inadequacies in water supply and delivery systems. It is also distinct from *aridity*, a permanent and stable natural condition under constant climate in a given region, and *desertification* which, although usually associated with drought or aridity, is attributable to significant change in the ecological regime of a region by human activities (see table 6.1) exacerbating the effects of low rainfall. Lack of water retention in degraded environments may have the same effect on agricultural production as lack of rainfall.

To define or measure drought in any kind of absolute terms is difficult, however many attempts have been made to do so — Chapman (1976) for example, lists thirty-seven ways of measuring drought in terms of moisture deficit, but Yevjevich et al. (1983) point out that the most meaningful depend on a property of the *run-length*, the duration of the

Table 6.1 Drought, aridity, and desertification

Drought A temporary phenomenon	Aridity A permanent condition	Desertification A human-induced condition
Can occur in any climate	Climatic state associated with global circulation patterns	Resulting from:
Unpredictable occurrence		Mining of groundwater
Uncertain frequency	Low annual precipitation normal	Overgrazing and unwise cultivation
Uncertain duration	Rainfall highly variable in both time and space	Attempts to extract from land more than natural productivity allows
Uncertain severity	High evaporation normal	
Long period of lower-than-average precipitation	High solar energy input	Unwise irrigation practices leading to salinisation
Protracted period of diminished water resources	Large annual temperature variations normal	and producing symptoms of:
		Reduction of perennial vegetation cover
Diminished productivity of natural ecosystems	Low productivity of natural ecosystems normal	Aquifer depletion, land subsidence
		Damaged surface soil and subsoil, loss of soil nutrients
Diminished productivity of farms and rangeland	Low productivity of farms and rangeland normal	Water and wind erosion of soil, decreased infiltration
		Increase in soil temperature
Deterioration of farmland and rangeland	Sparse human settlement normal	Compaction and salinisation of soils
		Oxidisation of soil organic matter
		Reduction of water-holding capacity of soil
		Increased propensity for flash flooding and further erosion
		Loss of productivity of natural ecosystems and of farms and rangeland
		Raised surface albedo (tends to diminish rainfall)
		Invasion of former farms and rangeland by woody weeds

peiiod that effective precipitation is below some specified value, or of the *run-sum*, the accumulated moisture deficit. The wide variation in normal precipitation between different areas of a country the size of Australia or the USA requires that drought be characterised in terms of local, or at the very least, regional climate. And moisture deficiency is defined not merely by lack of rainfall, but also by the amount of evapotranspiration, runoff, and infiltration. Because all drought definitions must ultimately be tied to particular systems of resource use, there has resulted much disagreement over the amount, duration and extent of moisture deficiency necessary to establish a drought threshold within specific areas. In addition, because drought occurs with varying frequency in all regions of the globe, in all types of economic systems, and in developed and less developed countries alike, the approaches taken to define drought reflect differences in ideological perspectives as much as regional differences.

Because drought is a 'creeping phenomenon', accurate prediction of either its onset or end is a difficult task. In Australia, the Bureau of Meteorology has identified an association between drought and the El Niño–Southern Oscillation (ENSO) phenomenon. The Southern Oscillation is essentially a decrease in the normal atmospheric pressure gradient between the subtropical high pressure cell in the eastern south Pacific and the normally low pressure over the region north of Australia (Lockwood, 1984). It is now known to be associated with the El Niño phenomenon, which occurs quasi-periodically over several years, when warm tropical waters move southward along the western coast of South America, overlaying large areas of the cold Peruvian current with warmer water. The appearance of an El Niño is very often associated with below average rainfall over northern and eastern Australia, with most major droughts on record being related to marked negative swings in the Southern Oscillation Index, and the Bureau of Meteorology regularly uses ENSO data for seasonal drought prediction.

The Australian Bureau of Meteorology defines drought as a condition of *serious moisture deficiency* if the rainfall in an area falls in the first decile (i.e., the lowest 10 per cent of all previous totals for the same period of the year) for a period of three months or more. A condition of *severe moisture deficiency* is identified if rainfall is within the lowest 5 per cent of all previous totals for the same period of the year, for three months or more. The Bureau maintains a 'Drought Watch' to identify the onset of any unusually dry period over any part of the continent and, once identified, issues monthly information bulletins designed to assist decisions by farmers, graziers and government. Information is updated until arrival of adequate rainfall to 'break' the drought.

IMPACTS OF DROUGHT

During the settlement of the semi-arid zones of both Australia and USA, there was a general over-optimistic appraisal of moisture availability as a characteristic of the resource potential of areas, a perception which persisted at least until mid-twentieth century. Many droughts have affected small and large areas of Australia since settlement, and the average annual cost of drought to the Australian economy is over $300 million (AWRC, 1992). During the period when reasonable weather records have been kept, there have been severe droughts of national importance in 1864–66, 1880–86, 1888, 1895–1903, 1911–16, 1918–20, 1939–45, 1958–68 and 1982-83 (Bureau of Meteorology, 1989). The 1895–1903 event was especially disastrous for woolgrowers, as the onset of the drought came at a time of economic depression, and also coincided with the climax of the rabbit irruption in New South Wales. Over 50 million sheep died in Australia at the time (reducing the sheep population by more than half its pre-drought numbers), and cattle numbers were reduced almost by half (Foley, 1957; Gibbs and Maher, 1967). Coughlan et al. (1979) reported that there had been 31 regional droughts in the previous 30-year period, with rarely a year without drought in some part of the country. Some droughts persisted up to ten years. Record low rainfalls from April 1982 to February 1983 affected most of Australia, but the drought was especially severe over South Australia, Victoria, most of southern and inland New South Wales, and central and southern Queensland, where rainfall deficiencies were the lowest on record. Losses were estimated in excess of $3 billion not counting the losses from dust storms and the catastrophic bushfires of 16 February 1983 in Victoria, which were caused, at least in part, by the drought. Similarly, a broad area of serious to severe rainfall deficiency over much of Queensland and New South Wales in 1993 (with some parts having recorded lowest ever 18-month rainfall) contributed to the worst bushfires in New South Wales in living memory in January 1994.

In the USA, during the drought years of the 1930s associated with the infamous 'Dust Bowl', wheat yields fell by more than 32 per cent overall, and corn by 50 per cent. Low crop yields, and crop failures, year after year, led to the failure of about 200 000 farms and the migration of over 300 000 people from the southern Great Plains states. Information on the impact of drought periods on a worldwide basis is very difficult to collate. International aid agencies such as the Red Cross make estimates of numbers of people who have died or have been displaced by droughts for some periods and countries, but information comparable with that for other disasters is relatively scarce (UNEP, 1991).

What is the cost of drought? Is it the value of lost production, of the value of potential economic benefits which may have been attained in the absence of a drought? Is it the cost of mitigation measures which individual rural producers and governments employ to survive the drought? Is it the sum of both of the above? In terms of sheer human suffering, the impact of drought (famine) on millions of people, particularly in Africa, is largely lost upon a world desensitised to pictures of near-naked, emaciated children flashed on its television screens. Closer to home, for most readers, is the hardship and suffering of rural families caught in a negative spiral of increasing indebtedness as they attempt to maintain a viable farm enterprise in the face of both increased costs and lower returns (sometimes negative returns) engendered by drought. Bates (1976: 219) presents the doctrinaire economist's position:

It is only possible to consider drought as imposing an avoidable economic cost to the extent that the potential benefits from measures which could be implemented to reduce drought losses are greater than those which could be obtained from the alternative investments which would have to be forgone to implement the measures concerned.

The issues are complex, and make any kind of meaningful cost–benefit analysis of the value of drought mitigation programs (as one might carry out for a flood mitigation program, say) extremely difficult, if not impossible. For example, at the level of the individual enterprise, a grazier may be faced with the prospect of selling a substantial proportion of his stock, on a very depressed market in drought, for a trivial return, only to be faced with the costs of restocking at some time later, when the drought has broken. On the other hand, should he choose to retain the stock, in the hope that conditions will improve, he may be faced with a huge fodder bill (plus interest). In the former case, returns from the productivity of the animals are forgone, but direct costs associated with management of them are reduced: in the second, the management costs remain, but the productivity of the animals in a drought may be very low, even with supplemental feeding.

Rarely is more than 20 per cent of Australia affected by a drought at any one time. The effects on the national economy are greatest when drought occurs in southeastern Australia. The direct impacts are, principally, failure of crops, fall in livestock numbers, and increased slaughterings, leading to decline in farm income, fall or elimination of the liquid assets of the farmer, decline in farm investment expenditure, rise in the level of rural debt, and degradation of land (see table 6.2). Impact of a drought may be somewhat lagged in the case of livestock production, but is immediate on rain-watered crops. Stock losses rarely exceed 20 per cent of the national total and 40 per cent of a regional total: there is usually only a small rise in food prices and in Australia few go hungry since

Table 6.2 The impact of drought

Economic	Environmental
Stock production	Fish
Impaired productivity of rangeland	Damage to fish habitat
Forced reduction of stock	Insufficient flows for fish
High cost/unavailability of water for stock	Loss of young fish due to decreased flow
High cost/unavailability of feed for stock	Animals
Grassland fires more likely	Damage to wildlife habitat
Crop production	Lack of feed and drinking water
Failure of annual crops	Vulnerability to predation or disease (e.g., from concentration near water)
Damage to perennial crops: crop loss	Plants
Impaired productivity of cropland (from wind erosion, etc.)	Impaired productivity of ecosystems
Insect infestation	Insect infestation more damaging
Wildlife damage to crops	Wildlife damage to ecosystems
Timber production	Wildfire more likely
Wildfire more likely	Water quality
Tree disease	Salt concentration increased
Insect infestation	Air quality
Impaired productivity of forestland	Dust pollution
Fisheries production	Aesthetic quality of landscape
Damage to fish habitat	Dust
Insufficient flows for fish	Vegetation dead or dying
Loss of young fish due to decreased flow	*Social*
Businesses	Public safety
Decreased purchasing power in rural areas	From forest and grassfires
Increased customer demand for credit	Public-health related
Increase in bankruptcies (farm & town)	Diminished sewage dilution, and increased pollutant concentrations (low river flows)
Industries dependent on agricultural production (fertiliser manufacturers, food processors)	Aesthetic
Decline in viability	Personal hygiene diminished, domestic gardens suffer, and dirty cars and streets if water restrictions severe
Governments	Unemployment
Revenue losses from reduced tax base	Esp. in drought-related production
Cost of water transport or transfer	Loss of ownership due to foreclosures
Cost of development of new water sources	Loss of savings
	Uncertainty

domestic food production is usually well in excess of demand. Most of the population, being urban, is insulated from the effects of drought. Land degradation, however, can be serious with widespread erosion and over-grazing in cropping and pastoral lands.

The significance of drought in the national economy has declined since the nineteenth century, as the proportion of GNP represented by rural production has declined, the transport systems have improved, and because growth in the non-farm sector is not necessarily affected by drought. Nevertheless, drought does cause major production losses and drastic personal hardship in specific regions of Australia. Rainfall often accounts for 60 per cent or more of the variation in agricultural produc-tion. Costs associated with this production do not vary greatly with rainfall, and hence net farm income falls dramatically in a drought and can be negative. These variations may cause a multiplier effect of two or threefold on the output and income of the affected region. A drought therefore affects the business life of small communities and creates unem-ployment, and it is for the above reasons that Australian federal government in the past made finance available for drought relief.

There may have been some positive outcomes from drought, in Aus-tralia at least. A study of the land laws of New South Wales (ably summarised for the period to 1956 by King, 1957), and of the reports of various commissions of inquiry into the state of arid and semi-arid land management (the latest being in 1984, see Fisher, 1984) reveals that droughts have been significant in stimulating changes in land manage-ment and resource use in Australia. Heathcote (1983) also draws attention to the benefits to certain sectors (e.g., transport) that may result from demands of drought mitigation, and the fact that public works (such as road improvements) having long-term benefits may be stimu-lated by demands of drought relief.

Dust storms

Dust storms originate from wind erosion of fine soil particles, often during drought, or accompanying desertification. Wind erosion of farmland degrades the soil and can cause other damage such as burial and/or abra-sion of crops and infrastructure, and 'clogging' of transport, farm and domestic equipment. The long-term soil loss reduces fertility of the soil, its water holding capacity and its ability to withstand further erosion. The health and well-being of people and animals also can be seriously affected by dust. The effects include lung, eye and allergy problems, and health problems may be exacerbated by the drying out and blowing of particles from salt lake beds during drought years. Dust storms have been reported

Figure 6.1 Dust storm over Melbourne in 1983. The clouds of dust were a secondary product of the severe 1982–83 drought in eastern Australia. Photograph courtesy Bureau of Meteorology, Australia.

from Australia, Asia, North America and Africa (Pye, 1987; Heathcote, 1983; Wheaton, 1992). Stories such as Steinbeck's *The Grapes of Wrath* may seem to place dust storms in the past, but they still occur (figure 6.1), and effects are numerous, severe, serious and costly. The conditions leading to dust storms often have been rightly blamed on human activities, primarily the cultivation of erosion-sensitive soils in semi-arid areas, such as the marginal Australian wheatlands, or the 'Dust Bowl' of the USA (see, e.g., Heathcote, 1983). Although it is true that unwise land use in many parts of the world accelerates wind erosion, dust storms are not totally due to human intervention: they are also a natural phenomenon in all low rainfall regions. Dust storms have occurred throughout the earth's history: deposits of loess, wind-carried, silt-sized material, can be found in many regions of the world. The distribution and frequency of dust storms globally has been considered by Goudie (1978, 1983) and for specific sites by Changery (1983), Guedalia et al. (1984), Hagen and Woodroffe (1973), Nickling and Brazel (1984), Orgill and Sehmel (1976), and Wheaton and Chakravarti (1987, 1990).

MANAGEMENT OF DROUGHT

The impacts of drought on society and on the environment often linger for years after the drought has passed, and actions taken during non-drought periods may greatly influence the level of vulnerability to a subsequent drought episode. Drought planning is a dynamic, iterative process (Wilhite and Easterling, 1989), which requires post-drought evaluation of the success (or otherwise) of mitigation measures, and planning for the next drought, which will surely arrive, by preparing for or managing drought risks as part of the routine course of business. The very first drought mitigation strategy on record, dated about 1500 BC, emphasised forward planning and drought preparedness — it is recorded in the book of *Genesis* how Joseph, governor of Egypt, directed that stockpiles of grain were to be assembled as a drought management strategy.

Today, maintenance of fodder reserves (and/or other forms of insurance, if available), and fine tuning of agricultural practices to optimise water use (summer fallow, strip farming, ploughless fallow, tillage practices that reduce moisture loss by weeds, runoff agriculture, microcatchment agriculture), and numerous other strategies (table 6.3) are commonplace among primary producers in technologically advanced countries. Government research institutes and universities also carry out research and development on pro-active drought management measures, including precise definition of crop requirements for water, genetic manipulation of crop plants, use of halophytic crops, and introduction of crops adapted to dry conditions (e.g., rubber from guayale, seed oil from jojoba, resin from gumweed — Hinman, 1984). However, there is no management strategy that can contain all drought impacts: even with the best management practices, farmers/graziers can at best only *minimise losses*, not prevent them, and catastrophic drought, with failure of the farm enterprise and bankruptcy, can affect the best of farm managers.

Institutional drought mitigation has usually been characterised by reactive, rather than pro-active, management strategies, most frequently involving emergency provision of fodder or water to affected areas. Such measures attract much media attention, and the publicity profile makes it easy for government to satisfy the public that 'something is being done'. Of more importance (Wilhite and Easterling, 1989) are clearly focused drought contingency plans, prepared well in advance by government, with the aim of early identification of drought, lessening of personal hardship, improvement of the efficiency of resource allocation, and the ultimate reduction of drought-related impacts and the very need for government involvement in relief programs. Referring to the African drought crisis of the 1980s, Mayer (1985: 4688) observed that

The devastating famine that is striking African nations from Mauritania to Ethiopia is being treated as a sudden disaster. In fact, an early warning system . . .

Table 6.3 Drought mitigation measures

Water supply management	Water demand management	Mitigation of drought impact
Use existing supplies	Efficiencies in water use	Insurance (if available)
Surface storages	Legal restrictions on water use	Supplemental feeding
Subsurface storages	Public pressure for conservative use	Progressive reduction in stock numbers
Enlargement of conveyance capacities	Economic disincentives on use	Relief measures
Deepening of wells or bores	Recycling of water by user	Concessional loans (low interest rate,
Interbasin transfer	Decrease of production	sometimes with deferred starting date
Long distance transport from	Of water–using commodities	for repayments)
unaffected areas (road, rail)	Of high water demand crops,	Restocking loans to primary producers
Temporary reduction of conservation	then of other crops	Loans for purchase of fodder
flows	Migration of stock	Loans for supply of water
	Transport of stock to second property	Carry-on loans for business
Develop new supplies	Agistment of stock	Subsidies
Induce aquifer recharge	Migration by nomads	On freight charges for stock movement
Emergency use of lakes	Inhibition of evaporation or	On freight charges for carriage of
Desalination of seawater or brackish	evapotranspiration (use of sealant layers	water to primary producers or local or
water	or coverings on water storages, stubble	semi-government authorities
Use of weather modification for	mulching, artificial mulch materials)	Stock slaughter or disposal subsidies
rainfall or snowpack augmentation	Reduction of seepage losses	Well drilling subsidies
Use of fossil waters	In irrigation canals	Agistment subsidies
Snow and ice management	In urban water supply networks	
Dew and fog harvesting		
Use of low-quality supplies (short-term		
measure)		
Relaxation of water quality standards		

could have been in place [based on] meteorological data . . . satellite photography . . . economic data on prices and stores of basic foods in famine prone regions, and health data . . . on children in vulnerable areas.

Shears (1980) also commented on the fact that agricultural rehabilitation measures, desperately required in some of the poorer countries to raise the level of drought preparedness, are not given a priority by international agencies concerned with drought (famine) relief simply because the management of relief programs is dominated by health professionals — their assessment of the problem and their recommended responses tend to be in terms of health and nutrition parameters.

For a government-sponsored drought relief plan to be viable, reliable information must be rapidly disseminated in order that producers are provided with adequate information about drought, alternative management strategies and assistance measures available; and there must be appropriate impact assessment techniques, especially in the agricultural sector, for use by government to identify periods of abnormal risk and to allow for timely initiation of assistance measures. Drought declaration procedures must be well publicised and consistently applied. The responses and attitudes of landholders may depend somewhat on the timing, availability and quality of government drought relief, with some even avoiding early management decisions until government's intentions are known. Assistance provided under a national or state drought policy should not discourage agricultural producers or other water-dependent entities from themselves adopting appropriate and efficient management practices that alleviate the effects of drought — assistance measures must not discriminate against good farm managers, by 'rewarding' those who have failed to adopt appropriate individual drought risk-management strategies. Assistance policies must also emphasise the importance of protecting the long-term viability of the natural and agricultural resource base (Wilhite, 1986). Policies which incorporate the above ideals have been developed in South Africa: while Australia has made drought-stricken farmers ineligible for disaster relief, South Africa has introduced a system of relief which encourages farmers to undertake sustainable practices that also reduce the likelihood of agricultural drought (Smith, 1993).

In Australia, state governments are responsible for providing advisory services to landholders on all aspects of farming, including land use and soil conservation, and are responsible for carrying out applied research into local problems. While the basis for drought relief finance varies from state to state, in general, finance has been sometimes made available to landholders during a drought for a wide range of measures such as freight concessions on fodder supplies and stock removal, agistment costs, provision of seed wheat and fertiliser, carry-on loans, and for restocking after the drought. The main aim appears to have been to enable landholders to stay on their properties, in the hope that conditions will improve. In effect, the drought relief measures may only help restore the *status quo* to the farmlands, rather than emphasising development of land-use practices better adjusted to the realities of harsh climatic regimes (Heathcote, 1983).

However, over the years, landholders have been gradually adopting both short- and long-term measures to reduce the effects of drought. Short-term measures lead to decisions on whether to hand-feed sheep, to let them die, to transfer them on agistment or to sell, often on a falling market price. Emergency grazing on Travelling Stock Routes or road easements (known colloquially in Australia as 'the long paddock') may provide escape routes for starving stock. In general, landholders are reluctant to keep fodder reserves in excess of that required to carry livestock over the normal dry period of the year. It is costly to store, it deteriorates in quality during storage and it may not be sufficient to carry the livestock throughout the drought. Financial reserves may be more important in drought management of livestock than are fodder reserves, although extra conserved fodder may reduce the high cost of restocking after a drought. Proposals have been put forward from time to time that federal government should operate a national drought fodder reserve based on wheat and hay storages. Such schemes have never come to fruition, mainly because fodder conservation is regarded by government as a landholder's responsibility, the location of storage may not be near the location of the next drought, and the length of storage before being required (and hence the cost of storage) is unknown. The long-term measures adopted by landholders include provision of adequate watering points, increased fencing to permit deferred grazing, contour ploughing and stubble mulching to increase infiltration of rainfall, the use of more water-efficient crops and pastures, the purchase of a second property in a less drought-prone area and the creation of financial reserves.

Outside of the permanent large-scale irrigation schemes, there are only relatively small areas which are specifically irrigated to help a landholder carry through a drought. In central Australia, for example, although most properties have supplies of underground water sufficient for small-scale irrigation, the total irrigated area is very small. The main value of these irrigated areas is to provide cheaper fodder for cattle in transit. Few of the pastoralists have the expertise to manage an irrigation area and failures are common because of alternative needs for labour and finance, e.g.

cattle-yards and fences. In general, irrigation is not an economic means of coping with drought on Australian pastoral properties.

The water users most likely to be affected by drought in the future are the city dwellers with their need for water for industry, parks and gardens. Restrictions on water use over summer have already been imposed in some years in several Australian capital cities (Melbourne, Perth, Adelaide), while increasing levels of salinity in some river systems in drought years presents problems of corrosion in water pipes and industrial equipment.

Water, as a life support fluid, is priceless and has no substitute, but in general it is a relatively low-valued fluid for which neither the demand nor supply is very mobile. For uses such as maintenance of ecological values, transportation, hydroelectric power generation, recreation and waste disposal, the unit value of water is very small indeed, especially at the margin, Even in uses such as food production, or for most industrial processes, the value of an extra unit of water is small. Normal water management strategies must be designed on a low cost per unit basis, and few, if any nations have either the water or financial resources to be able to design and build supply systems with large excess capacity or network redundancies.

CONCLUSION

In the future, pressure on the marginal lands of the world for food and fibre production will not decrease. Maximising production from these lands, many of which are prone to the effects of drought, will be a priority, and the development of appropriate guidelines for their management, with cognizance that periodic drought is a normal system response and must be accommodated, is vital. We can no longer concentrate on reactive management responses which merely perpetuate the *status quo*, leaving the pre-conditions for further disaster in place, but must instead concentrate on development of sustainable management regimes. Nor can we afford a repetition of the naïve responses exemplified by the provision of extra watering points in the Sahel, which, rather than enabling optimal use of available feed by existing stock, produced the counter-intuitive result of overstocking, leading ultimately to greater drought-related stock losses and dreadful desertification (after all, cattle was a form of wealth). It is probably true that, over the years, identification of drought as primarily a physical problem of water balance and plant/animal system response has encouraged scientific research which has emphasised solutions in the physical, rather than the societal realm, and impeded development of drought management strategies with a holistic perspective, which address it as a meteorologic, hydrologic, agricultural and socioeconomic phenomenon. Drought is a complex phenomenon, with profound societal ramifications, and indeed, the ultimate measure of the severity of a drought is that of its impact upon society.

7

A management model for natural hazards

Disaster planning is usually carried out in a climate of public apathy and economic constraints, and can frequently be shown to have followed what Langton and Chapman (1983) referred to as a 'crisis-response' model. In Australia, for example, land management provisions for the western lands of New South Wales can be shown to have been stimulated by the disastrous drought and rabbit plague of 1895–1903, and coastal management legislation in Queensland stemmed from impact of tropical cyclones in 1967. History also shows that deaths and property damage from ocean storms and storm surge, in the low-lying coastal areas of Britain–NW Europe, were significantly reduced in each 50- to 100-year period following a major disaster which triggered a campaign to build up and/or restore sea defence barriers (Lamb, 1991), and effective sea defences for the Netherlands (possibly the most ambitious civil engineering project ever undertaken, relative to the small size of the nation) were constructed following catastrophic sea flooding in 1953.

Perception that there is a problem — awareness that death, injury, or property loss has reached an unacceptable level — and belief in our ability to manipulate the natural environment, or human interaction with that environment, are necessary precursors to hazard management by society (Palm, 1990). Societies with a fatalistic attitude are likely to respond passively to hazard: if people believe they can do nothing to reduce their susceptibility to hazard, they will not take action. And even where belief in human ability to control the environment is present, government attitudes may vary from *caveat emptor* ('let the buyer beware'), on the one hand, to one of assuming responsibility for the protection of citizens' lives or property, on the other. An appropriate position is not easy to define, since individual freedom and sovereignty is at stake. It may be argued that strict social controls may reduce risk for those who, left to themselves, are at risk from the unmitigated harshness of the environment, but there are few who would wish to pass on to government the responsibility for behaviour that they can undertake as individuals. Who would wish the government to be their keeper? It would seem that governmental action is warranted only in those cases where individual action is excessively costly, counter-productive, or even impossible.

Management analyses of natural hazards involve a number of clearly defined aspects, linked as shown in figure 7.1. These are:

- *Event analysis* of the natural hazard: establishing the probability of occurrence, within a specific period of time in a given area, of a potentially damaging natural phenomenon of given magnitude.
- *Vulnerability analysis*: assessment of the potential degree of loss to a given entity at risk, or set of such entities, resulting from the occurrence of a natural phenomenon of a given magnitude. The entities at risk are the population, buildings and civil engineering works, economic activities, public services, utilities and other infrastructure in a given area.
- *Risk analysis*. Risk is a function of the probability of the specified natural hazard event and vulnerability of cultural entities.
- *Response analysis*, which is concerned with investigating the strategies available to manage a particular hazard.
- *Decision analysis*: using objective criteria and rational methods of decision making to arrive at the most viable management option in any particular case.

EXTREME EVENT ANALYSIS

Extreme event analysis is crucial for natural hazard management. Without being able to answer the questions 'How big, and how often?', management of natural hazards cannot be placed upon a rational basis. If a very large number of measurements of a certain type of event, say, earthquake magnitudes, were available from a period of record at least as long as the planning period, the problem is relatively simple, since in that case the magnitude of the design event ('how big?') could be derived for probability levels of interest ('how often?') directly from the sample data.

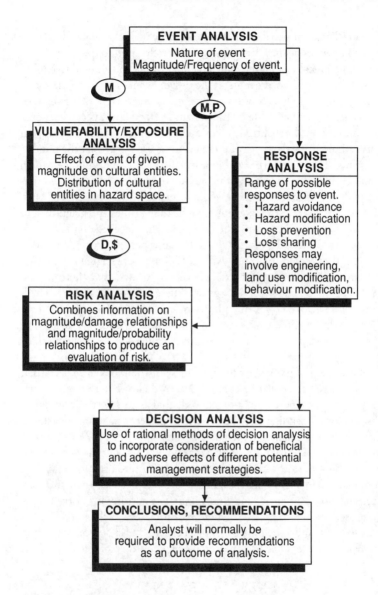

Figure 7.1   The process of natural hazard management.
M = Magnitude of natural event;
P = Probability of natural event;
D = Potential damages.

The ideal amount of data will almost certainly not be available, however, and a known frequency distribution is generally fitted to the sample data (either graphically or mathematically), and this in turn is used to extrapolate from the recorded events to the design event. The data base for an extreme event analysis consists of the most extreme event of a given type from each year of record (e.g., the maximum flood flow in each year on a certain river). When fitted by an appropriate frequency distribution curve, the statistical pattern of these events is used to predict the number of events of the given type, within a specified range of magnitude, which could be expected to occur during a discrete time period (e.g., the number of floods bigger than X that would be expected on a certain river in a 100-year period). Extreme event analysis thus defines the magnitude of the event which we can expect to be equalled or exceeded on the average once every N years, the N-year event. In a long period of record, say 1000 years, it is expected that there will be 1000/N events equal to or larger than the N-year event (for example, it is expected that there will be 1000/50 = 20 floods equal to or greater than the '50-year event'). These events are *random*, however, and *non-cyclic*. There is no implication that they will occur at more or less regular intervals of N years. The average period of N years between such events is called the Return Period, or Recurrence Interval.

Many different probability (or frequency) distributions have been used in hazard management (Bell et al., 1989). By their very nature, the distributions of extreme events are skewed, that is to say, there is tail of observations extending to either very high or very low values (mostly to the right, to high values, as seen in the example in figure 7.2), in contrast with the normal distribution, which is symmetrical about the mean. Continuous distributions, such as that in figure 7.2, are used to define the *magnitude* of an event corresponding to a given *probability of occurrence*. The *Lognormal*, *Gumbel*, and *Pearson* are the continuous distributions most used for hazard analysis. Discrete distributions such as the Poisson are used to describe the probability of a certain *number of events* occurring *within a given time frame* (for an example of a Poisson plot, see figure 3.4).

VULNERABILITY ANALYSIS

Vulnerability analysis asks: 'given a specified hazard type and intensity, what damages will accrue to specific cultural entities?' Vulnerability analysis requires an understanding of the response of types of structures and structural qualities to the specified hazard. The outcome of vulnerability analysis is an understanding of the relationship between different intensities of the specified hazard and resulting damages to cultural entities (cf. figure 7.3).

Exposure analysis is the spatial aspect of vulnerability analysis: 'what is at stake in a specific area?' It requires knowledge (i.e., inventory) of the distribution and type of structures and other property, and of people. The outcome of vulnerability/exposure analysis is some measure of loss (usually expressed in dollars or in loss of life) in relation to the different measures of intensity or magnitude of the hazard concerned. Geographic Information Systems (GIS) are of particular assistance in developing exposure analyses. GIS provide for storage, display and processing of spatial data. GIS allow us to account for variations of the parameters involved in a spatially distributed problem, and to summarise, integrate, and display complex spatial information for efficient hazard planning and decision making.

The costs of natural disasters are, however, more extensive than simply those which may be considered under the headings of damage or loss to a given element at risk. Cost, in broad terms, is measured by the costs of long-term human response as well as by the frequency/magnitude aspects of the physical agent and its potential for direct damage. Thus the influence of the hazard includes the costs of preparedness (scientific research, planning, management, engineering measures, opportunity costs, etc.), as

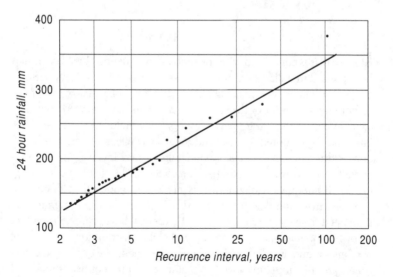

Figure 7.2 Frequency distribution of maximum 24 hour rainfall from each year of record at Turramurra, New South Wales, Australia, the wettest station in the Sydney region.

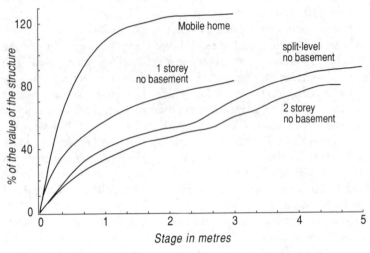

Figure 7.3   Typical damages to residential structures as a function of depth of flooding

well as secondary effects such a loss of production in industry and services, unemployment, and increased public health and social expenditures.

RISK ANALYSIS

Risk analysis is concerned with the probability of defined loss. Losses may be expressed in dollars or in terms of loss of life or injury. Ideally the outcome of risk analysis is the ability to specify the probability of loss of any magnitude as a result of the specified hazard within a defined area. The *direct* effects of the hazard event are normally included in risk analysis, but it should be noted that in total the *indirect* effects — the transfer of impact through social and economic links — may involve greater losses and affect many more people than the direct effects of the event (cf. figure 2). Sometimes a distinction is made between *risk*, defined as a situation in which the probability distributions for the outcomes are developed from a sound empirical or theoretical basis, and are therefore 'known', and *uncertainty*, when the probability distribution of the set of outcomes is undefined, there is no consensus among the experts, and outcomes are therefore 'unknown'. Accurate perception of risk is hindered by what has been termed 'the thirst for certitude' (Ruckelshaus, 1983): people in general prefer certainty rather than probability statements, and may reduce the anxiety associated with low-probability statistically defined events by denying their existence.

How safe is safe enough? Otway and Eardmann (1970) have made some generalisations about levels of acceptable risk associated with accidents in Western society:

- Accidents with a probability of about 10^{-6} (i.e., one in one million) deaths per person/year are not of great concern to the average person. The individual may be aware of them, but feels that they will never happen to him. Phrases associated with these occurrences express an element of resignation: 'Lightning never strikes the same place twice', 'An act of God'.

- Mortality risks at the level of 10^{-5} (1 in 100 000) per person/year are above the cognition threshold of ordinary people. Mothers warn their children about most of these hazards (playing with fire, drowning, firearms, poisons) and some people accept a degree of inconvenience, such as not travelling by air, to avoid them. Safety slogans for these risks have a precautionary ring, 'Never swim alone', 'Never point a gun at another person', 'Keep medicines out of children's reach'.

- At a fatal accident level of 10^{-4} (1 in 10 000) per person/year, people are willing to spend money, especially public money, to control the cause. Money is spent for traffic signs and control, and police and fire departments are maintained with public funds. Safety slogans popularised for accidents in this category show an element of fear, e.g. 'The life you save may be your own.'

- However, if a risk of fatal accident approaches a level of the order of 10^{-3} (1 in 1000) per person/year, immediate action is taken to reduce the hazard. This level of risk appears unacceptable to everyone.

Attitudes to death or loss from disaster are not neutral with respect to the cause or location of the event. The shooting of five or six people by a crazed gunman will be reported with horror on national media, while the death of double that number of people on the roads on a busy weekend may pass without comment. Five hundred people shot in civil strife in a foreign country will arouse less interest than five people killed by earthquake near home.

An unresolved question concerns the joint assignation of risk and responsibility within society. To what extent should an individual be held responsible if he or she chooses to live, for example, in a known bushfire-prone or flood-prone locality? To what extent is it appropriate to spend the funds of taxpayers generally in order to rescue, or protect the life or property, of those who choose to live in an area of known, and high, environmental risk? To what extent does the governing body have a responsibility to protect citizens from diffuse environmental hazards?

RESPONSE ANALYSIS

Response analysis is concerned with answering the question: 'what can be done?' In general, there are five (or possibly six) categories of response:

- *Avoid the hazard.* Most hazards have spatial limits, and some have temporal limits as well, so that it may be possible to reduce or eliminate risk by appropriate location of human activities in space (land-use planning) or in time (scheduling). Avoiding a hazard may involve the use of buffer zones (building setbacks) or other zoning to exclude vulnerable forms of land use from hazard-prone areas. The seasonal nature of some hazards may make it possible to program vulnerable activities so that they are not coincident with the peak hazard risk. For example:

    Flood — Locate activities on known flood-free sites; designate floodplain for land uses minimally disrupted by rare flood event, such as public open space or playing fields.

    Earthquake — Locate activities away from known faults.

- *Modify the causal factors of the hazard.* Change the characteristics of the natural event so that its impact on human activities is diminished or nullified. For example:

    Flood flow: Reduce and/or impede runoff by land management practices such as afforestation or contour banking.

    Earthquake: Lubrication of fault, by injecting water.

- *Modify the hazard environment.* For example:

    Flood: Impound flows in reservoir for later release, constrain flow to channel with levees, facilitate discharge by channel modification.

    Earthquake: Soil and slope stabilisation, tsunami barriers.

- *Modify loss potential.* Accept that the natural event will occur, and that we cannot control it, but take measures to minimise damage to or loss of property or life. Building codes or other measures to ensure that structures are appropriately engineered for the degree of hazard to which they are exposed may be employed. Regulations may be framed to preserve natural protective features such as sand dunes or forests. For example:

    Flood: Warning systems coupled with emergency flood-proofing and/or evacuation.

    Earthquake: Earthquake-proof or earthquake-resistant building design (including post-construction earthquake-proofing), disaster management procedures in place and tested, isolation devices for oil and gas lines, fire management procedures.

- *Share the losses.* Accept that the natural event will occur and that losses are inevitable, but attempt to minimise impact on individuals through loss sharing arrangements such as insurance. Loss sharing usually involves some sort of formal contractual arrangement such as insurance, but informal kinship or community links may also function very well. Loss sharing arrangements also involve government hand-outs to disaster victims, tax write-offs, or relief payments organised by charities or service clubs.
- *Do nothing.* Accept outcomes of hazard and bear the losses. This is not necessarily the apathetic option: it may be the most rational option in the light of considered analyses of risk. Bearing of one's own losses is the most cost-effective strategy where losses are trivial and/or any form of organised effort may cost more than the benefits produced. Unfortunately in many poor or poorly organised societies, loss bearing is often the only option available.

DECISION ANALYSIS

Decision analysis involves (1) decomposing and structuring the problem (often the least emphasised, but most critical phase of the decision process), (2) assessing the values and uncertainties of the possible outcomes, and (3) determining the preferred strategy in terms of some specified choice criteria.

Decision analysis separates the roles of the executive decision maker, the hazard expert, and the analyst. The analyst's role is to structure a complex problem in a tractable manner so that the consequences of the alternative actions may be assessed. Various hazard experts provide the technical information from which the analysis is fashioned, but it is the decision maker who acts for society in providing the basis for choosing among the alternatives. The analysis provides a mechanism for integration and communication so that the technical judgments of the experts and the value judgments of the decision maker may be seen in relation to each other, examined, and debated. Decision analysis makes not only the decision but the decision process itself a matter of formal record. For any complex decision that may affect the lives of thousands, a decision analysis showing explicitly the uncertainties and decision criteria can and should be carried out.

In determining the preferred management strategy, the fundamental aim is to ensure that the desirable outcomes (or *benefits*) from a chosen strategy are considerably greater than the undesirable outcomes (or *costs*). It is obvious that the expected benefits from any project should exceed the costs, otherwise there would be no justification for the project to proceed. Where both costs and benefits may be clearly expressed in dollar values, cost–benefit analysis provides a useful basis for decision making. In natural hazard management, the 'benefits' provided by any management strategy are usually defined as the future losses avoided by implementing the strategy. These losses may only be averaged out over a period of years (the 'planning period'), and cost–benefit analysis provides us with a mechanism for comparing the immediate costs of implementing a strategy with the probable future losses avoided. Simple cost–benefit analysis alone is, however, rarely adequate as a basis for decisions about hazard management. Quantification of many of the variables is very difficult. Some analysts have equated the value of a human life with presumed earning power, for example, but any conversion of human life or death, suffering or well-being, or environmental degradation or improvement, to dollar values must at best be a subjective process.

Related issues are concerned with the varying emphases that individuals and groups in society place on aspects of the outcomes from any project, and with who pays the costs and who reaps the benefits from any form of environmental management. Clearly, decision models capable of handling a great variety of social and environmental data, and of displaying such information in a form which aids rational choice among alternative courses of action, are required. The complexity of the problem necessitates the use of some form of multi-criteria analysis, and microcomputer-based models (e.g., LUPIS (Cocks and Ive, 1988; Chapman, 1992b), Micro-QUALIFLEX (Ancot, 1989), or DEFINITE (Janssen, 1993)) are available to evaluate interactively the impact of the introduction or omission of different decision criteria, and of applying different weights to criteria — the latter may be used to incorporate the 'votes' of affected groups in society. Multi-criteria decision models have the advantage of being able to incorporate strictly quantitative criteria (for example, dollar costs associated with number of dwellings destroyed, or protective engineering) as well as those which are difficult to quantify, but for which it is often easy to develop an ordinal ranking system (e.g., 'Strategy C has minimal impact on environmental quality, strategy B maximal, and strategy A ranks somewhere in between'), and combining them into an overall preference specification for the alternative strategies under consideration.

AN EXAMPLE APPLICATION

The urge to possess a piece, even if only a small share, of beachfront property has often led people to build in hazardous situations at the seashore.

Once the hazard is perceived, remedial measures such as the construction of seawalls may lead to the destruction of the beach amenity and opportunities for recreation which provided the reason for settlement in the first place. Management of the coastal erosion hazard (see 'Storm waves and coastal erosion' in chapter 3), like other problems which arise where human activities come into conflict with natural processes, is complex, and involves conflicting objectives. In this case, the principal objectives may be identified as protection of property from shore erosion and maintaining the amenity values of the beach zone (while at the same time minimising the costs of management actions), but there are others, such as the perceived safety of beachfront residents, for example. Let us examine how the management model outlined above might be applied to a problem of coastal erosion hazard management at a beach we shall call Narragon. The discussion which follows is largely based on a case reported by Chapman (1991). In the case under analysis, the beach is neither receding nor advancing over the long term but is subject to probabilistic erosion: the hazard is from periodic wave destruction of expensive beachfront dwellings and infrastructure, either from direct wave impact or from wave erosion of the frontal dune on which the buildings rest (see figure 3.7). There is little risk to human life, as people usually have the opportunity to leave a structure before it collapses.

The first step in applying the management model is *event analysis*: establishing the probability of occurrence of the natural phenomenon over the potentially damaging range of magnitudes. Here it is important to establish a measure of magnitude which is directly related to the impact of the hazard upon human affairs. For floods we use depth of inundation or discharge, for example, and for earthquakes there are familiar Richter magnitude or Mercalli intensity measurements. When it comes to coastal erosion, we could think of measures of storm wave height, or power, or surge effects, but there is one single parameter which is directly related to property damage, and that is the amount by which the frontal dune is eroded, measured as recession from a datum (such as the foredune toe), of the erosional 'scarp' which forms under storm wave attack. Measurements were made on Narragon beach in the field, following storm events, and from both ground and aerial photographs, to compile a database of the *frequency* with which different *amounts* of erosional cut occurred. These data were then fitted by a statistical *frequency distribution* (figure 7.4) to allow us to specify the probability of any given amount of dune recession. Following common planning practice, we will plan for the 100-year Recurrence Interval event, and consider costs over a 50-year planning period.

We now turn our attention to the system subject to the hazard, to *vulnerability analysis,* which in this case is the assessment of the potential amount of loss to beachfront houses and infrastructure resulting from the occurrence of any given amount of dune recession. If cut were small, damages would mostly be restricted to fences, gardens, and street ends, with associated utilities. More critically, however, we are concerned with establishing the amount of recession which would cause 10 per cent, 20 per cent . . . 50 per cent, etc., damage to the average shorefront property, in a similar way to the flood damage relationship shown in figure 7.3. Exposure analysis is then concerned with presenting this information in a way which is useful to planners and decision makers, often in a map-based form. In the case of the coastal erosion hazard here, it is convenient to express the result of vulnerability/exposure analysis as the relationship between amount of erosion and dollar loss per kilometre of shorefront (figure 7.5).

We are now in a position to talk about risk. *Risk analysis* is concerned with the probability of defined loss, and is a function of the probability of the specified natural hazard event and the vulnerability of cultural entities. In our vulnerability analysis, we defined the dollar losses associated

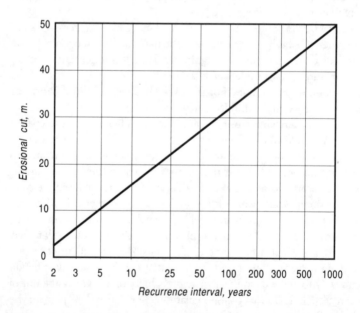

Figure 7.4 Probability of erosional cut at Narragon beach, shown on a Gumbel frequency distribution plot. Erosional cut measured from dune toe inland. Probability expressed as Recurrence Interval for defined amount of cut.

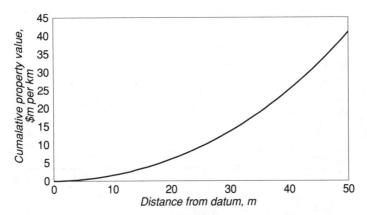

Figure 7.5 Value of property at risk from coastal erosion at Narragon, expressed as cumulative value from dune toe inland, per kilometre of shoreline.

with different amounts of shore erosion, and in the event analysis, we assessed the probability of different amounts of shore erosion. Clearly, we can combine these results to give us the risk factor, which is the *probability of dollar losses* over the range of possible values (figure 7.6). The area under the curve shown in figure 7.6 is the statistically 'expected' annual loss or, in, other words, the probable average annual loss due to the hazard concerned. It is a vital piece of information for management purposes because, if we succeed in eliminating the hazard as a result of some management strategy, the annual losses which we have prevented give us a measure of the benefits that our management strategy has produced. As we are planning for the 100-year event (probability = 0.01), we take the area under the curve which includes the 100-year event and all events which are more common, which in this case is $975 644.

Before making any planning decisions however, we must concern ourselves with *response analysis*, with investigating the strategies available to manage a particular hazard. Let us suppose for this example that only three management alternatives are available:

1 To construct a revetment, which is a substantial loose-boulder wall extending high enough to prevent wave overtopping, and keyed into the beach below erosion base level. A properly designed revetment would virtually eliminate any possibility of future property damage, but significantly reduces the amenity value of the shore zone. Cost (original construction plus maintenance over 50-year period) is estimated at $6.86 m per kilometre.

2 Beach nourishment, which by adding sand to the beach sand prism, enhances its role as a buffer for storm wave attack. Any sand surplus to

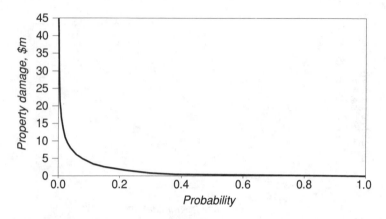

Figure 7.6 Risk of property loss from coastal erosion at Narragon. There is a high probability (close to 1.0) of very small losses, and a very low probability (close to 0.0) of very large losses. Values per kilometre of shoreline.

normal wave demand will accumulate at the dune toe, and the beach management program will be organised to stabilise that sand with vegetation, thus improving both erosion resistance and aesthetic appeal of the beach. Beach nourishment is not a once-for-all management option, and will require 'topping up' from time to time. Cost (initial application plus maintenance applications over 50-year period) is estimated at $5.3 m per kilometre. Risk is not completely removed, however, as the nourished beach is a buffer, not a barrier, to erosion.

3 To take no action, which is always an option in hazard management, but in addition, the estimation of the effects of taking no action provides a datum from which to measure the changes caused by the other options. There are no direct management costs, but the amenity value of the beach declines over time.

We are now ready for *decision analysis*, the use of objective criteria and rational methods of decision making to arrive at the most viable management option. If the problem were purely one of dollar losses and dollar costs, we could employ simple *cost–benefit* analysis, but in this case some of the beneficial and adverse effects of the alternatives arise from things that are not traded in markets and are therefore not readily susceptible to dollar evaluation. Let us start by examining the cost–benefit scenario, and go on to see how the problem might be set up for *multi-criteria analysis*.

How cost–benefit analysis works

In natural hazard management, the 'benefits' provided by any management strategy are defined as the future losses avoided by implementing that strategy. As these losses may be considered in terms of a statistically 'expected' *annual loss*, cost–benefit analysis provides us with a mechanism for comparing the immediate costs of implementing a strategy with the probable future losses avoided. To compare future losses with expenditures made in the present, a technique known as *discounting* is used. The discount rate may be likened to an inverse interest rate. It works this way:

In the case under analysis, the probable average annual loss due to coastal erosion is $975 644. Let us assume that an annual interest rate of 7 per cent prevailed. The formula for evaluating the value of an investment at compound interest is $P(1 + r)^n$, where P is the principal, or initial amount invested, r is the interest rate, and n is the number of years the investment is to run. Conversely, the amount P which should be invested now, in order to return an amount X in n years time, at an interest rate of r, is found by: $P = X/(1+r)^n$. Therefore, we could prepare for a loss of $975 644 in year 20, say, by investing $252 125 now at 7 per cent interest. Consequently we can say that the *net present value* of a $975 644 loss expected in year 20 is $252 125 if discount rate is 7 per cent. Similarly the net present value of a specified loss in each year up to the limit of our planning horizon could be evaluated, and the total of these amounts would then represent the net present value of losses expected from the hazard. In our example, for a hazard with 'expected' losses of $975 644 per year, for a 50-year planning period:

Year 1	$975 644/1.07	= $911 817
Year 5	$975 644/1.07^5	= $695 621
Year 20	$975 644/1.07^{20}	= $252 125
Year 50	$975 644/1.07^{50}	= $ 33 121
	Total for 50 years	= $ 13.46 m

If the present value of the damages bill from a hazard is $13.46 m, it clearly does not make sense to spend more than that amount to provide protection from the hazard. However, if protection from the 100-year event could be provided for, say, $6.86 m, which is the cost of the revetment, it could be argued that a benefit (in the form of future losses avoided) of $13.46 m has been acquired at a cost of $6.86 m thus providing a ratio of benefits to costs (the *benefit–cost ratio*) of 13.46/6.86 = 1.96. In making decisions on the basis of cost–benefit analysis alone, the benefit–cost ratio must be significantly greater than 1.0 for a project to be viable and, in the case of two competing strategies, the one with the highest benefit–cost ratio would be favoured.

How multi-criteria analysis works

The major goals of environmental management are economic efficiency, environmental quality and social well-being. The first may be evaluated

Table 7.1 Planning balance sheet, Narragon beach erosion management
(Costs per kilometre of beach, over 50-year planning period, discounted @ 7%)

Criteria	Alternative management strategies		
	1 Revetment	2 Nourishment	3 No action
(a) Minimise total costs	Expected losses = $13.46 m Cost of revetment = $6.86 m Benefit = $6.6 m	Expected losses = $13.46 m Cost of nourishment = $5.3 m Benefit = $8.16 m	Expected losses = $13.46 m
(b) Maximise aesthetic appeal of beach	Structure ugly Rank = 3	Vast improvement Rank = 1	Declines over time as residents install makeshift erosion control measures Rank = 2
(c) Maximise recreational amenity of beach	Structure interferes with beach access, and may interfere with beach processes, reducing recreational amenity Rank = 3	Improved beach and dune quality Rank = 1	Declines over time as residents install makeshift erosion control measures Rank = 2
(d) Maximise perceived safety of beachfront residents	Residual risk trivial Rank = 1	Residual risk present Rank = 2	Erosion risk not reduced at all Rank = 3
Mean rank, criteria b-d:	Overall rank = 2.33	Overall rank = 1.33	Overall rank = 2.33

by dollar criteria, such as cost–benefit analysis; the latter two may not. In the case of Narragon beach, we identified the principal management objectives as protection of property from shore erosion (an economic efficiency objective), and maintaining the amenity values of the beach zone, which has aspects both of environmental quality (e.g., aesthetic appeal of beach) and social well-being (e.g., recreational amenity, perceived safety of beachfront residents). Perception of safety by residents depends on confidence felt about the efficiency of engineering works, or, in their absence, seems to be a function of beach width and dune–beach system stability. These objectives are by no means the only ones that could be chosen, but they seem to be a reasonable set for the purpose of demonstration. Effective analysis of a multi-criteria problem demands a computer model (e.g., Micro-QUALIFLEX (Ancot, 1989), or DEFINITE (Janssen, 1993)), but a simplified presentation of the method may be made by setting out the alternative courses of action, and the criteria upon which decisions are to be made, in the form of a planning balance sheet, as shown in table 7.1.

Disregarding the no-action alternative, in the Narragon example a choice would have to be made between two alternatives whose net costs do not differ much. The decision would depend on the weights the decision maker(s) might give to the relevant aspects of environmental quality and social well-being. On the face of it, the nourishment alternative, with highest overall rank at 1.33, and greatest net benefit, may appear to be the clear winner, but if the decision maker considered the minimisation of residual risk the most important single non-dollar criterion, the revetment would be favoured.

CONCLUSION

The process of natural hazard management is complex, as hazards frequently interact, and the management strategies appropriate for a specific hazard at a location may conflict with other land management goals. Despite our technological prowess, there are still uncertainties inherent in all cases where we attempt to modify the behaviour of natural processes, and perhaps even more uncertainties where we attempt to influence the reactions of society. There are always secondary effects associated with any natural disaster, and in total these may exceed the primary impacts in terms of dollar costs, or, in some cases, benefits, as there are always those who gain in some way from a natural disaster.

Conclusion —
the outlook

We live in the information age. We have remarkable monitoring and predictive capabilities for natural hazards, we regularly employ advanced telecommunications and simulation modelling in emergency management, and we exploit the potential of geographic information systems (GIS) in hazard mitigation. New developments in information technology can improve disaster management by analysing data virtually as fast as they are collected. The scenario of digital weather stations relaying data to a computer model to predict behaviour of a wildfire which has just been reported by thermal sensing from satellite, with the output from the simulation model input to an expert system to develop preferred fire management tactics, that are then relayed in map form via mobile fax to fire crews already on their way to the scene, for example, is not speculation but technological reality.

And yet, loss of life and property from natural disaster continues to rise. Why is this so?

For a start, world population has been rising at an unprecedented rate in the twentieth century, and continues to rise rapidly: there will be far more people and property in the twenty-first century for natural hazards to affect than today (Quarantelli, 1993). In addition, the pressure of population on land resources will mean that many more people will live and work where the risks of disasters are highest (e.g., in river valleys and on coastal lowlands). Many of the poorest people will have nowhere to settle but in the areas of greatest risk, such as steep slopes and flood-prone areas near the expanding fringes of mega-cities in the developing world, areas which in many cases have been avoided in the past precisely because they are hazard prone. Some natural disasters are also being exacerbated inadvertently by human activities consequent upon population pressure, such as deforestation of mountain areas leading to intensification of lowland floods. And there is also the problem of changing climate and the rise in sea level that is expected to accompany it.

The so-called 'greenhouse', or global warming scenario is, *per se*, slow moving (although pervasive), but some of the outcomes postulated from it certainly have far-reaching implications for natural hazard management. On a shorter time scale, periodic astronomical phenomena such as sunspot or tidal cycles, or quasi-periodic climatic phenomena such as the El Niño–Southern Oscillation (ENSO) phenomenon have been implicated in the occurrence of a variety of natural hazards, and sudden, or catastrophic, changes to the atmosphere, as from a large explosive volcanic eruption, may involve complex subsidiary interactions affecting all of the globe.

Changes in the content of CO_2 and other trace gases (especially methane (CH_4), nitrous oxide (NO_2) and the chlorofluorocarbons ($Cl_n Fl_n$)) in the atmosphere are thought to be responsible for global warming and sea level rise (Harrington, 1987; Pearman, 1988; Titus, 1988; Smith and Tirpak, 1988; Ramanathan, 1988; Barth and Titus, 1984). Much of the current excess of CO_2 in the atmosphere has been released through the burning of fossil fuels, but there is disequilibrium in the global carbon cycle due to deforestation and the consequent reduction of the carbon store in the forest biomass (Sedjo, 1989). The principal impacts of climatic change on sea level rise appear to be thermal expansion of the oceans as a result of global warming (although the coupling mechanisms and lag times involved are not known with precision); increase of ocean volume resulting from glacier retreat and ice-sheet melting/collapse; and changes in major oceanic circulation structures (gyres, currents). There is some consensus of possible sea level rise of 0.8 m ± 0.6 m over the period 1990–2040 (see, e.g., Pearman, 1988). As sea level rises, drowning of coastal lowlands and coastal erosion may be anticipated — substantial shoreface, beach and dune erosion may occur as a result of quite modest sea level rise, with the eroded sediment being redistributed over the lower shoreface (Bruun 1962, 1983). But the greatest threat is to low-lying coastal zones, such as the Netherlands, about half of which is below present high-tide level. The Dutch are therefore acutely conscious of the level of the ocean beyond their sea defences, and have already evaluated the costs of protecting against a rise of 1 m in sea level. Assuming that defence is the policy option chosen, rather than retreat or abandonment, a program of strengthening sea defences to keep pace with the predicted sea level rise over the next century would cost about $US3 000 000 000 (in comparison, cost of maintenance is presently about $US30 000 000 per year), and adaptation of harbours, locks, and bridges, etc., to cope with a 1 m sea level rise would cost about $US500 000 000.

Climatic change assumes particular importance in connection with the frequency of extreme events. Berz (1988), commenting on the dramatic increase in the impact of natural disasters on the insurance industry in the last few decades, noted that the problem 'will be dramatically aggravated if the greenhouse predictions come true'. An outcome of the 'greenhouse' scenario is likely to be increases in the frequency of extremes (Blong, 1992; Mitchell and Eriksen, 1993) — post-climatic change frequency of events defined as extreme events in terms of present levels of human adaptation might change quite markedly as a result of even a small change in the mean values of important climatic variables. The significance of extreme events following one another in quick succession lies in their cumulative impact, especially on those living on the margins of agricultural viability. A single extreme event may be withstood, but if buffer stocks of food were depleted by one very bad season, a second in succession could be devastating. In the future, global warming is expected to aggravate the vulnerability of areas subject to storm surge, storm wave attack and windstorm caused by tropical cyclones. The global warming scenario involves the possibility of higher frequency of tropical and extratropical cyclones, increased intensity of tropical cyclones, and an underlying rise of sea level, on which any surge would be superimposed. Sensitivity analysis by Love (1988) indicated that, of the three possibilities, the sea level rise is the most certain, significantly shortening recurrence intervals for what are now regarded as extreme surge events. Love concluded that increased cyclogenesis is less likely, although it is suggested that, for a rise in sea-surface temperature of 1°C, an additional 2–3 tropical cyclones per year may be expected to form in the Australian region (Love,1988). Higher sea surface temperatures may also result in more intense tropical cyclones, significantly increasing expected surge heights — Murty and El-Sabh (1992) point out that the higher sea surface temperatures postulated by some climatic change scenarios may produce tropical cyclones in the Bay of Bengal with central pressure lower than 850 mb (in comparison, the tropical cyclone producing the devastating storm surge of 1970 in Bangladesh had a central pressure of 940 mb).

It is well known that there are eleven and twenty-two year cycles in sunspot activity, and there are cycles in the tides. *Spring tides,* when tidal range is greater than average, for example, appear at intervals of about two weeks. These are caused by the alignment of the sun, moon and earth on approximately a straight line, a relationship known as syzygy which occurs twice in each period of 29.53 days. There are other cyclic movements in the relative positions of the heavenly bodies which mean that the complete sequence of tidal variations takes 18.6 years. The inter-relationship of the solar orbit of the earth, and the earthly orbit of the moon, plus dynamic influences involved, occasionally bring the three bodies unusually close together at or near time of syzygy, the *syzygy-proxigee-perigee* alignment, producing *perigean spring tides.* Wood (1976), in a study of about 100 coastal storms occurring between 1635 and 1976, drew attention to the association between the occurrence of perigean spring tides and major coastal storms in the North American region, and it is perhaps noteworthy that 1974, a year of particularly close perigee-syzygy alignments, was also a year of major coastal erosion damage in Australia (Chapman et al., 1982). Associations have also been made between solar and tidal cycles and episodes of storminess, floods, and drought (Chapman et al., 1982; Currie, 1981, 1984).

The El Niño–Southern Oscillation phenomenon (ENSO) has certainly been implicated in extremes of behaviour of many natural hazard phenomena: floods and drought in India (Bhalme et al., 1983), North America (Glantz, 1984), and Australia (Bureau of Meteorology, 1989); tropical cyclone frequency (Gray, 1984; Holland et al., 1988); wind-storms (Horel and Wallace, 1981); and severity of sea ice (Marko et al., 1988), to name but a few.

Infrequently, but almost certainly, we will have to confront the potential effects of a sudden and marked increase in stratospheric aerosols. This was presented in the Cold War era as a potential outcome of nuclear war (the 'Nuclear Winter' scenario), but natural occurrences have been associated from time to time with large explosive volcanic eruptions. The 1883 eruption of the Indonesian volcano Krakatau introduced about 20 km³ of ash into the atmosphere, the event being implicated in a drop of worldwide average temperature by about 0.5°C for some years afterwards. Larger explosive eruptions have occurred in historical times — the eruption of Tambora (also in Indonesia) in 1815 is estimated to have released 150–180 km³ of ash into the atmosphere, and although meteorological recordings were almost non-existent at the time, it is known that 1816 was called 'the year without a summer' in North America and Europe. A medium term (few years) temperature drop need not be at the lethal freezing level in order to be catastrophic, as, for every organism, low temperature thresholds exist which can, in decreasing order of severity, cause: death; weakening to the point of losing out in the struggle for existence; loss of resistance to infection; inhibition of reproduction; or decline/disappearance of food sources.

But that is not all. Despite our technological sophistication, it is not at all apparent that we have culturally adapted to the realities of natural hazards. Scientific and technological knowledge is vital for effective hazard mitigation, management, and disaster response. But to what extent is responsibility for vulnerability to natural events attributable to deficiencies in the political-economic-social structure, not directly related to our scientific and technological capability for hazard response? There is a very real sense in which environmental risk may be considered to be primarily a function of the value systems of a society, of the fabric of aspirations, beliefs, myths and ways of living and acting as they are expressed on the level of a common mentality, and it has long been realised that decisions about whether to mitigate a natural hazard are most often not a function of how dangerous the hazard is in absolute or objective terms but how dangerous it is perceived to be (see, e.g., White, 1974). Quantification, so beloved of science, may not necessarily lead to a more rational depiction of reality (Miller, 1985). Collective culture has the ability to create, enhance or ameliorate vulnerability — Alexander (1991) has argued that the overdevelopment of hazardous sites has become a virtually unstoppable force in the Western world because capital, expertise and investment are heavily concentrated into perpetuating the phenomenon and that to alter such a situation would lead to economic recession.

However, a completely risk-free society is not a realistic goal, and in any event, there are few who would wish to live in the mothball-and-padded cell type of environment which that concept invokes. In technologically advanced societies, we individually have much less exposure to natural hazards than our forebears, but on the other hand have gladly accepted the hazards inherent in the comforts of life that technology provides, and learned to live with those hazards. All responsible parents inculcate in their children, even as toddlers, a healthy respect for electricity, for example, and deaths from misuse of that most versatile form of energy are extremely rare. We value the personal freedom and convenience of the motor car so highly that we are prepared to accept (in most cases gladly) the very considerable constraints on behaviour, including conformist driver education, associated with the widespread use of those devices. And for those few who persist in 'learning by accident', we have little sympathy.

In the technologically advanced societies, we have long since accepted zoning controls designed to preserve environmental quality (e.g., location of certain types of industry in relation to residential areas), and more recently utilised the zoning concept for hazard mitigation (e.g., controls on building within the 100-year floodline on floodplains). We have evolved building codes for the purpose of ensuring minimal standards of structural integrity and environmental quality, and have extended these to cover some kinds of hazard management (e.g., earthquake-resistant construction is required in specified areas of seismic risk). But we also cherish the locational mobility and freedom of choice which allows us to live in an area of high environmental amenity, such as a peri-urban bushland setting, that also carries with it a high degree of risk from environmental hazard, such as bushfire. If the individual takes such choices fully cognizant of the risks involved, and is quite prepared to accept the consequences, there can surely be no argument against such a course of action. No one would desire the Orwellian nightmare of full state control over the way we live. Yet experience has shown that individuals do place themselves and their property at risk through ignorance, because of faulty perception ('It's never happened here in living memory, so we're safe', 'You cannot prevent an act of God'), and through building designs which maximise exposure to environmental amenity but also increase vulnerability to environmental hazard. And when disaster does strike, there are usually very considerable demands on community resources for emergency response and subsequent rehabilitation, along with acrimony from citizens relating to perceived sins of technological omission, committed by political and technical authorities who 'failed' to protect constituencies from the vagaries of nature. There is clearly room for improvement, and this will come through social-economic-political means, in minimising exposure to hazard by tightening the nexus between choice and responsibility, between action and consequences, between knowledge and behaviour.

In the less-developed world, appalling death and injury tolls from natural disasters may appear to be more related to lack of material resources than to deficiencies in the political-economic-social structure. But is that really so? Watt (1983), for example, using the example of Nigerian drought, argued that natural hazards are 'not simply natural', for though a drought may be a 'catalyst' for famine, 'the crisis itself is more a reflection of the ability of the socio-economic system to cope with the unusual harshness of ecological conditions and their effects.' State hazard responses along lines which are apparently technically sound may be quite counter-productive — Bernard (1985), for example, shows that government strategies for the development of the Kenyan drylands actually worsened the drought hazard by undermining traditional responses to environmental hazards. In some cases propensity for hazard, even in the presence of appropriate countermeasures, is attributable to social factors unrelated to the natural event — Matsuda (1993), for example, shows that even where storm surge shelters were available in Bangladesh, refuge-seeking behaviour was hampered by (1) illiteracy: some people could not understand the widespread and simple instruction signs; (2) the imperfect cadastre: because actual location, limits, and ownership of many parcels of land was unknown or ill-defined, occupation was indispensable to claim right of ownership, so people were reluctant to leave their land; and (3) cultural inhibition on close physical contact between the sexes which prevented many females from taking refuge in shelters where population densities reached at least five or six people per square metre. Sadly, also, in many parts of the less-developed world, civil strife and political forces have disrupted the functioning of the socioeconomic system to the extent that natural hazard crises become natural disasters — the tragedies in Africa are only too well known.

Finally, the last word has not yet been written on the management of natural hazards. The statement made by Burton and colleagues a generation ago (Burton et al., 1963) is as true today:

A paradox is presented in man's apparent growing susceptibility to injury from natural hazards during a period of enlarged capacity to manipulate nature ... Nature retreats on every hand and man, armed with a burgeoning technology, is asserting his ecological dominance more surely. Nevertheless, in every month the mass media continue to report in dramatic fashion the occurrence of natural disasters.

■

References

Alexander, D. E. (1991) 'Applied Geomorphology and the Impact of Natural Hazards on the Built Environment'. *Natural Hazards*, 4: 57–80.

Alexander, M. E. (1982) 'Calculating and interpreting forest fire intensities'. *Can. J. of Botany*, 60: 349–57.

Ancot, J. P. (1989) *Micro-QUALIFLEX*. Kluwer Academic Publishers, Dordrecht.

Andrews, C. J., Cooper, M. A., Darveniza, M., and Mackerras, D. (eds) (1992) *Lightning Injuries: Electrical, Medical, and Legal Aspects*. CRC Press, Inc., Boca Raton, FL.

Arnell, N. W. (1987) 'Flood insurance and floodplain management'. In J. Handmer (ed.), *Flood hazard management: British and international perspectives* (pp. 117–33). Geo Books, Norwich.

Austin, A., and Heather, N. (1988) *Australian arachnology*. Spec. Pub. 5, Aust. Entomol. Soc.

AWRC (1992) *Floodplain management in Australia*, 2 vols. AGPS, Canberra, for Australian Water Resources Council.

Bagdonas, A., Georg, J. C., and Gerber, J. F. (1978) *Techniques of frost prediction and methods of frost and cold protection*. Technical Note No. 157, World Meteorological Organisation, Geneva, 160pp.

Baker, V. R., and Costa, J. E. (1987) 'Flood Power'. In L. Mayer and D. Nash (eds), *Catastrophic Flooding*. Allen and Unwin, Boston.

Baker, V. R., and Pickup, G. (1987) 'Flood geomorphology of Katherine Gorge, Australia'. *Geol. Soc. Am. Bull.*, 98: 635–46.

Barker, D., and Miller, D. (1990) 'Hurricane Gilbert: anthropomorphising a natural disaster'. *Area*, 22(2):107–16.

Barth, M. C., and Titus, J. G. (1984) *Greenhouse Effect and Sea Level Rise*. Van Nostrand Rheinhold, New York.

Bass, D. J. (1991) 'White Cypress Pine pollen: an important seasonal allergen in rural Australia'. *Med. J. Australia*, 155: 572–7.

Bass, D. J., and Baldo, B. A. (1990) 'The spectra of pollen allergens from six common grasses'. *Rev. Paleobotany & Palynology'*. 64: 87–95.

Bates, W. R. (1976) 'National economic effects of drought in Australia'. In T. G. Chapman (ed.), *Drought* (pp. 217–42). AGPS, Canberra.

Baxter, P. J. (1990) 'Medical effects of volcanic eruptions'. *Bull. of Volcanology*, 52: 532–44.

Bell, F. C., Dolman, G. S., and Khu, J. F. (1989) 'Frequency Estimation of Natural Hazards and Extremes'. *Aust. Geogr. Studies*, 27: 67–86.

Berg, R. L., and Wright, E. A. (1984), *Frost action and its control*. American Society of Civil Engineers, New York, 151pp.

Bernard, F. E. (1985) 'Planning and environmental risks in Kenyan drylands'. *Geographical Review*, 75: 58–70.

Berz, G. (1988) 'Climatic change: impact on international reinsurance'. In G. I. Pearman (ed.), *Greenhouse: planning for climate change* (pp. 579–87). CSIRO, Melbourne.

Berz, G. A. (1991) 'Global warming and the insurance industry'. *Nature and Resources*, 27(1): 19–24.

Bhalme, H. N., Mooley, D. A., and Jadhav, S. K. (1983) 'Fluctuations in the flood/drought area over India and relationships with the Southern Oscillation'. *Monthly Weather Review*, 111: 86–94.

Bilham, R. (1988) 'Earthquakes and urban growth'. *Nature*, 336: 625–6.

Bilham R. (1991) 'Earthquakes and Sea Level: space and terrestrial metrology on a changing planet'. *Reviews of Geophysics*, 29(1):1–29.

Blong, R. J. (1992) *Impact of climate change on severe weather hazards in Australia'*. AGPS, Canberra, for Dept. of Arts, Sport, Environment, Territories, 35pp.

Bolt, B. A. (1991) 'Balance of risks and benefits in preparation for earthquakes'. *Science*, 251 (11 January): 169–74.

Booth, B. (1979) 'Assessing volcanic risk'. *J. Geol. Soc. Lond.*, 136: 331–40.

Borunov, A. K., Koshkariov, A. V., and Kandelaki, V. V. (1991) 'Geoecological consequences of the 1988 Spitak earthquake (Armenia)'. *Mountain Research and Development*, 11: 19–35.

Bradford, J. (1993) 'Biological hazards and emergency management'. *Natural Hazards Observer*, 17(5): 1–3.

Bradshaw, L. S., Deeming, J. E., Burgan, R. E., and Cohen, J. D. (1983) *The 1978 National Fire-Danger Rating System: technical documentation*. General Technical Report INT-69, USDA Forest Service, 44pp.

Brown, C. A., and Graham, W. J. (1988) 'Assessing the threat to life from dam failure'. *Water Resources Bull.*, 24(6): 1303–09.

Bruun, P. (1962) 'Sea level rise as a cause of shore erosion'. *Proc. J. Waterways and Harbors Division*, American Society of Civil Engineers, 88: 117–30.

Bruun, P. (1983) 'Review of conditions for use of Bruun rule of erosion'. *Coastal Eng.*, 7: 77–89.

Bryceson, K. P., and Wright, D. E. (1986) 'An analysis of the 1984 locust plague in Australia using multitemporal Landsat multispectral data and a simulation model of locust development'. *Agricult. Ecosys. and Environment*, 16: 87–102.

Buechley, R. W., van Bruggen, J., and Truppi, L. E. (1972) 'Heat Island = death island.' *Environmental Research,* 5: 85-92.

Bureau of Meteorology (1977) *Report on cyclone Tracy, December, 1974.* AGPS, Canberra, for Bureau of Meteorology.

Bureau of Meteorology (1985) *A report on the flash floods in the Sydney metropolitan area over the period 5 to 9 November, 1984.* Bureau of Meteorology, Sydney.

Bureau of Meteorology (1989) *Drought in Australia.* AGPS, Canberra, for Bureau of Meteorology.

Burton, I., Kates, R. W., and White, G. F. (1968) *The Human Ecology of Extreme Geophysical Events.* Natural Hazard Research Working Paper No. 1, Department of Geography, University of Toronto, Toronto.

Carrara, A., (1991) 'GIS techniques and statistical models in evaluating landslide hazard'. *Earth Surface Processes and Landforms,* 16: 427–45.

Carrega, P. (1991) 'A Meteorological Index of Forest Fire Hazard in Mediterranean France'. *Int. J. Wildland Fire,* 1(2): 79–86.

Carter, A. O., Millson, M. E., and Allen, D. E. (1989) 'Epidemiologic study of deaths and injuries due to tornadoes'. *Amer. J. Epidemiology,* 130: 1209–18.

Casadevall, T. J. (1991) *First International Symposium on Volcanic Ash and Aviation Safety: Program and Abstracts.* Circular 1065, US Dept. of the Interior, Geological Survey, Reston, Va.

Chamberlain, E. R., Doube, L., Milne, G., Rolls, M., and Western, J. S. (1981a) *The experience of Cyclone Tracy.* AGPS, Canberra.

Chamberlain, E. R., Hartshorn A. E., Mugglestone, H., and Short, P., Svensson, H., and Western, J. S. (1981b) *Queensland flood report, Australia Day, 1974.* AGPS, Canberra.

Changery, M. J. (1983) *A dust climatology of the western United States.* NOAA, Asheville, NC.

Chapman, D. M. (1980) 'Coastal erosion management in Australia'. In W. L. Edge (ed.), *Coastal Zone 80* (pp. 2218–35). American Society of Civil Engineers, & US Office of Coastal Zone Management, New York.

Chapman, D. M. (1981) 'Coastal erosion and the sediment budget'. *Coastal Eng.,* 4: 207–27.

Chapman, D. M., (1991) 'Shore protection — conflicting objectives in decision making'. In O. T. Magoon, L. T. Tobin, H. Converse, V. Tippie, and D. Clarke (eds), *Coastal Zone 91* (pp. 2340–53). American Society of Civil Engineers, New York.

Chapman, D. M. (1992a) ASH FRIDAY — A bushfire management simulation model. Software program (121k) and documentation (36pp). Environmental Education Unit, University of Sydney, Sydney.

Chapman, D. M. (1992b) 'Information Management in Beach Planning'. *Coastal Management,* 20: 203–17.

Chapman, D. M., Geary, M., Roy, P. S., and Thom, B. G. (1982) *Coastal Evolution and Coastal Erosion in New South Wales.* Coastal Council of NSW, Sydney.

Chapman, T. G. (ed.) (1976) *Drought.* AGPS, Canberra.

Charpin, D., Vervloet, D., and Charpin, J. (1988) 'Epidemiology of asthma in western Europe'. *Allergy,* 43: 481–92.

Clarke, J. E., and Bach, W. (1971) 'Comparison of the comfort conditions in different urban and suburban microenvironments'. *Internat. Jour. Biometeorol.* 15: 41–54.

Coates, D. R., and Vitek, J. D. (1980) *Thresholds in Geomorphology.* Allen and Unwin, London. 498pp.

Cocks, K. D., and Ive, J. R. (1988) 'Evaluating a computer package for planning public lands in New South Wales'. *J. Environ. Mgmt,* 25: 249–60.

Cook, R. J., Barron, J. C., Papendick, R. I., and Williams, G. J. (1981) 'Impact on agriculture of the Mount St Helens eruptions'. *Science,* 211: 16–22.

Cornell, J. (1976) *The great international disaster book.* Charles Scribner's Sons, New York.

Coughlan, M. J., Hounam, C. E., and Maher, J. V. (1979) 'The drought hazard in Australia'. In R. L. Heathcote and B. G. Thom (eds), *Natural Hazards in Australia* (pp. 51-71). Australian Academy of Science, Canberra.

Covacevich, J., Davie, P., and Pearn, J. (1987) *Toxic plants and animals.* Queensland Museum, Brisbane.

Crozier, M. (1986) *Landslides, causes, consequences and environment.* Croom Helm, London.

Cunningham, C. J. (1984) 'Recurring natural fire hazards: a case study of the Blue Mountains, New South Wales, Australia'. *Applied Geography,* 4: 5–57.

Currie, R. G. (1981) 'Evidence of 18.6 year signal in temperature and drought conditions in N. America since 1800 AD'. *J. of Geophysical Research,* 86: 11055–64.

Currie, R. G. (1984) 'Periodic (18.6 year) and cyclic (11 year) induced drought and flood in western North America'. *J. of Geophysical Research,* 89(D5): 7215–30.

Dale, P. (1992) '1992 Mosquito management survey'. *Bull. Mosquito Control Assoc. Australia,* 4(2): 8–17.

Dale, P. (1993) 'Australian wetlands and mosquito control — contain the pest and sustain the environment?' *Wetlands* (Australia), 12: 1–12.

Davis, J. R., Hoare, J. R. L., and Nanninga, P. M. (1986) 'Developing a fire management expert system for Kakadu National Park, Australia'. *J. Environ. Mgmt,* 22(3): pp. 215–27.

Degg, M. (1992) 'Natural disasters: recent trends and future prospects'. *Geography,* 77: 198–209.

Delft Hydraulics Laboratory (1970) *Gold Coast, Queensland, Australia — Coastal Erosion and Related Problems* (2 vols). Report 257. Delft Hydraulics Laboratory, Delft, Netherlands.

Deshpande, B. G. (1987) *Earthquakes, animals and man.* Maharashtra Assoc. for the Cultivation of Science, Pune.

Donaldson, R. C. (1980) 'Geological hazards in Tasmania'. *Rec. Geol. Survey Qld,* 37: 97–118.

Dunkerley, D. L. (1976), 'A study of long term slope stability in the Sydney Basin, Australia'. *Engineering Geology,* 10: 1–12.

Elsom, D. (1989) 'Learn to live with lightning'. *New Scientist,* 122 (1670): 54–8.

Emmi, P. C., and Horton, C. A. (1993) 'A GIS based assessment of earthquake property damage and casualty risk: Salt Lake County, Utah'. *Earthquake Spectra,* 9(1): 11–33.

Erikson, K. T. (1979) *In the Wake of the Flood*. George Allen and Unwin, London.

Faupel, C. E., and Styles, S. P. (1993) 'Disaster education, household prepared-ness, and stress responses following Hurricane Hugo'. *Environment & Behavior*, 25: 228–249.

Fisher, M. C., (1984a,b,c,d) *First, second, third, and fourth Reports of the Joint Select Committee of the Legislative Council and Legislative Assembly to enquire into the Western Division of New South Wales*. Parliament of New South Wales, Sydney.

Fisher, R. V., Smith, A. L., and Roobol, M. J. (1980) 'Destruction of St Pierre, Martinique, by ash-cloud surges, 1902'. *Geology*, 8: 472–6.

Flannigan, M. D., and Harrington, J. B. (1988) 'A study of the relation of met-eorological variables to monthly provincial area burned by wildfire in Canada'. *J. Appl. Meteorology*, 27: 441–52.

Fleeton, M. W. (1980) 'Public and private adjustment to the bushfire hazard in Australia; empirical evidence from NSW'. *Australian Geographer*, 14(6): 350–9.

Foley, J. C. (1957) *Droughts in Australia*. Bull. No. 43, Bureau of Meteorology, Canberra.

Ford, D. T. (1981) 'Interactive nonstructural flood-control planning'. *J. of the Water Resources Planning and Management Division*, American Society of Civil Engineers, 107(WR2), Proc. Paper 16572, pp. 351–63.

Fountain, D. W., and Cornford, C. A. (1991) 'Aerobiology and Allergenicity of *Pinus radiata D. Don.* in New Zealand'. *Grana.*, 30: 71–5.

Fournier d'Albe, E. M. (1980) 'Approach to Earthquake Risk Management'. *Nucl. Saf.*, 21(2): 205–14.

Frank, N. L., and Husain, S. A. (1971) 'The deadliest cyclone in history?' *Bull. Amer. Met. Soc.*, 52: 438–44.

French, S. P., (1984) 'Applying Earthquake Risk Analysis Techniques to Land Use Planning'. *J. Am. Plann. Assoc.*, 50: 509–22.

Geipel, R. (1991) *Long-Term Consequences of Disasters: The Reconstruction of Friuli, Italy, in Its International Context 1976–1988*. Springer-Verlag, New York.

Gibbs, W. J., and Maher, J. V. (1967) *Rainfall deciles as drought indicators*. Bull. No. 48, Melbourne, Bureau of Meteorology, Melbourne.

Gill, A. M. (1975) 'Fire and the Australian flora: a review'. *Australian Forestry*, 38: 4–25.

Glantz, M. H. (1984) 'Floods, fires and famine: is El Niño to blame?' *Oceanus*, 27(2): 14–20.

Glickman, T. S., Golding, D., and Silverman, E. D. (1992) *Acts of God and Acts of Man*. Discussion Paper CRM 92–02, Resources for the Future, Washington, DC.

Goldammer, J. G. (1988) 'Rural land-use and wildfires in the tropics', *Agroforestry Sys.*, 6: 235–52.

Goudie, A. S. (1978) 'Dust storms and their geomorphological implications'. *J. Arid Environ.*, 1: 291–310.

Goudie, A. S. (1983) 'Dust storms in space and time'. *Progress in Physical Geography*, 7(4): 502–30.

Gray, M. (1992) 'Funnel-Webs: separating fact from fiction'. *Aust. Nat. Hist.*, 24: 32–9.

Gray, W. M. (1984) 'Atlantic seasonal hurricane frequency'. *Monthly Weather Review*, 112: 1649–67.

Guedalia, D., Estournel, C., and Vehil, R. (1984) 'Effects of Sahel dust layers upon nocturnal cooling of the atmosphere (ECLATS Experiment)'. *J. Climate Meteorol.*, 23: 644–50.

Gupta, H. (1992) *Reservoir induced earthquakes*. Elsevier, London.

Hadlington, P., and Gerozisis, J. (1985) *Urban pest control in Australia*. University of New South Wales Press, Kensington, NSW.

Hagen, L. J., and Woodroffe, N. P. (1973) 'Air pollution from dust storms in the great plains'. *Atmos. Environ.*, 7: 323–32.

Harrington, J. B. (1987) 'Climatic change: a review of causes'. *Can. J. For. Res.* 17: 1313–39.

Harris, S. L. (1990) *Agents of Chaos*. Mountain Press Publishing Company, Missoula, Montana, 260pp.

Heathcote, R. L. (1983) *The Arid Lands: Their Use and Abuse*. Longman, London.

Heathcote, R. L., and Thom, B. G. (eds) (1979) *Natural Hazards in Australia*. Australian Academy of Science, Canberra.

Hemmens, V. J., Baldo, B. A., Bass, D. J., Vik, H., Florvaag, E., and Elsayed, S. (1988a) 'A comparison of the antigenic and allergenic components of Birch and Alder pollens in Scandinavia and Australia'. *Int. Archs. Aller. Appl. Immunol.*, 85: 27–37.

Hemmens, V. J., Baldo, B. A., Bass, D. J., Vik, H., Florvaag, E., and Elsayed, S. (1988b) 'Allergic response to Birch and Alder pollen allergens influenced by geographic location of allergic subjects'. *Int. Archs. Aller. Appl. Immunol.*, 87: 321–8.

Hermes, N. (1987) *Crocodiles: Killers in the Wild*. Child & Assoc., French's Forest, NSW.

Hewitt, K. (1983) 'The idea of calamity in a technocratic age'. In K. Hewitt (ed.), *Interpretations of Calamity*. Allen and Unwin, Boston.

Hewitt, K. (1984) 'Ecotonal settlement and natural hazards in mountain regions: the case of earthquake risk', *Mountain Research and Development*, 4: 31–7.

Hinman, C.W. (1984) 'New Crops for Arid Lands'. *Science*, 225:1445–8.

Hirshleifer, J. (1987) *Economic behaviour in adversity*. Wheatsheaf Books Ltd, Sussex.

Hobbs, J. E., and Lawson, S. (1982) 'The tropical cyclone threat to the Queensland Gold Coast'. *Applied Geography*, 2: 207–19.

Holland, G. J., McBride, J. L., and Nicholls, N. (1988) 'Australian tropical cyclones and the greenhouse effect'. In G. I. Pearman (ed.), *Greenhouse — Planning for Climate Change*. CSIRO, Melbourne.

Hollingsworth, P. C. (1982) 'Landslides and residential development'. In *Landslide hazards in hillside development* (pp. 1-17). Proc. Seminar Engineering Geology Group, Geological Society of Australia, Griffith University, Brisbane, Nov. 1982.

Hopley, D. (1974) 'Australian Weather Example No. 2: Storm surge'. *Australian Geographer*, 12: 462–8.

Horel, J. D., and Wallace, J. M. (1981) 'Planetary scale atmospheric phenomena associated with the Southern Oscillation'. *Monthly Weather Review*, 109: 813–29.

Iida, J., and Iwasaki, T. (eds) (1983) *Tsunamis: their science and engineering*. Sendai Ofunato Kamaishi, Japan.

Ingles, O. G. (1974) *Unstable landforms in Australia*. Report No. 42. Aust. Water Research Foundation, Sydney.

Janssen, R. (1993) *Multi-objective decision support for environmental management*. Kluwer, Dordrecht.

JCR (1991) 'Impacts of Hurricane Hugo: September 10–22, 1989'. *J. of Coastal Research*, Special Issue no. 8 (Spring).

Johnson, E. A. (1992) *Fire and Vegetation Dynamics. Studies from the North American Boreal Forest*. Cambridge University Press, Cambridge, 129pp.

Johnson, R. W., and Threlfall, N. A. (1985) *Volcano town: the 1937–43 Rabaul eruptions*. Robert Brown and Assoc., Bathurst, NSW.

Jones, P. (1981) 'The geography of Dutch elm disease in Britain'. *Transactions of the Institute of British Geographers*, New Series, 6: 324–36.

Joyce, E. B. (1979) 'Landslide hazards in Victoria'. In R. L. Heathcote and B. G. Thom (eds), *Natural Hazards in Australia* (pp. 234–47). Australian Academy of Science, Canberra.

Kachroo, R. K., (1992) 'River flow forecasting'. *J. Hydrology*, 133: 1–178.

Katsura, J. (1992) *Storm surge and severe wind disasters caused by the 1991 cyclone in Bangladesh*. Japanese Group for the Study of Natural Disaster Science, Nagoyer, Japan.

Kaufman, Y. J., Setzer, A., Justice, C., Tucker, C. J., and Fung, I. (1990) 'Remote sensing of biomass burning in the tropics'. In J. G. Goldammer (ed.), *Fires in the Tropical Biota* (pp. 371–99). Ecological Studies 84, Springer-Verlag, Berlin.

Kawashima, K., and T. Kanoh (1990) 'Evaluation of indirect economic effects caused by the 1983 Nihonkai-Chubu, Japan, earthquake'. *Earthquake Spectra*, 6: 739–56.

Khalil, G. M. D. (1992) 'Cyclones and storm surges in Bangladesh: some mitigative measures'. *Natural Hazards*, 6: 11–24.

Kiil, A. D., and Grigel, J. E. (1969) *The May 1968 forest conflagrations in central Alberta — a review of fire weather, fuels and fire behaviour*. Information Report A-X-24, Canadian Forestry Service, Northern Forest Research Centre.

Killian, H. (1980) *Cold and frost injuries*. Springer-Verlag, Berlin.

King, C. J. (1957) 'An outline of closer settlement in New South Wales. Part 1 — the sequence of the Land Laws, 1788–1956'. *Rev. Mktg. & Agric. Econ.*, 25: 1–290.

Kreimer, A., and Munasinghe, M. (1991) *Managing Natural Disasters and the Environment*. The World Bank, Washington, DC.

Lack, D. (1954) *The natural regulation of animal numbers*. Clarendon, Oxford, 344pp.

Lamb, H. (1991) *Historic Storms of the North Sea, British Isles and Northwest Europe*. Cambridge University Press, Cambridge.

Langaas, S. (1992) 'Temporal and Spatial Distribution of Savanna Fires in Senegal and The Gambia, West Africa, 1989–90, Derived from Multi-temporal AVHRR Night Images'. *Int. J. Wildland Fire*, 2(1): 21–36.

Langton, B. J., and Chapman, D. M. (1983) 'The role of crisis in coastal erosion management in New South Wales, Australia'. *Coastal Zone 83* (pp. 1640–50). American Society of Civil Engineers, New York.

Leicester, R. H., and Reardon, G. F. (1979) 'Case studies of wind damage to domestic buildings'. In Heathcote, R. L., and Thom, B. G. (eds) *Natural Hazards in Australia* (pp. 204–15). Australian Academy of Science, Canberra.

Lewis, N. D., and Mayer, J. D. (1988) 'Disease as natural hazard'. *Progress in Human Geography*, 12(1): 15–33.

Lockwood, J. G. (1984) 'The Southern Oscillation and El Niño'. *Progress in Physical Geography*, 8: 102–10.

Lourensz, R. S. (1981) *Tropical cyclones in the Australian Region, July 1909 to June 1980.* AGPS, Canberra, for Bureau of Meteorology.

Love, G. (1988) 'Cyclone storm surges: post greenhouse'. In G. I. Pearman (ed.), *Greenhouse — Planning for Climate Change.* CSIRO, Melbourne.

Luke, R. H., and McArthur, A. G. (1986) *Bushfires in Australia.* Department of Primary Industry/CSIRO Division of Forest Research, Canberra.

Malingreau, J. P. (1990) 'The contribution of remote sensing to the global monitoring of fires in tropical and subtropical ecosystems'. In J. G. Goldammer (ed.), *Fires in the Tropical Biota* (pp. 337–70). Ecological Studies 84, Springer-Verlag, Berlin.

Malingreau, J. P., and Tucker, C. J. (1988) 'Large-scale deforestation in the Southeastern Amazon Basin of Brazil'. *Ambio*, 17: 49–55.

Marko, J. R., Fissel, D. B., and Miller, J. D. (1988) 'Iceberg movement prediction off the Canadian east coast'. In M. I. El-Sabh and T. S. Murty (eds), *Natural and man-made hazards* (pp. 435–62). Reidel, Dordrecht.

Marsh, T. J., and Monkhouse, R. A. (1993) 'Drought in the United Kingdom, 1988–1992'. *Weather*, 48(1): 15–22.

Matsuda, I. (1993) 'Loss of human lives induced by the cyclone of 20–30 April, 1991, in Bangladesh'. *GeoJournal*, 31(4): 319–25.

Matthews, A., and Jessell, M. (1993) 'Firescan'. *GIS User*, 3: 48–51.

Mayer, J. (1985) 'Preventing Famine'. *Science*, 227: 4688.

Mears, A. I. (1992) *Snow avalanche hazard analysis for land-use planning and engineering.* Colorado Geological Survey, Denver.

Miller, A. (1985) 'Technological thinking: Its impact on environmental management'. *J. Environ. Manage.*, 9: 179–90.

Miller, H. C. (1990) *Hurricane Hugo: Learning from South Carolina.* US Dept of Commerce, National Oceanic and Atmospheric Administration, National Ocean Service, Office of Ocean and Coastal Resources Management, Washington, DC.

Minor, J. E., and Mehta, K. C. (1979) 'Wind damage observations and implications'. *J. of the Structural Division*, ASCE, 105(ST11): 2279–91.

Mirtschin, P., and Davis, R. (1982) *Dangerous snakes of Australia.* Rigby, Adelaide.

Mitchell, J. K., and Eriksen, N. J. (1993) Effects of climatic change on weather related disasters. In J. Mintzer (ed.), *Confronting Climatic Change* (pp. 141–51). Cambridge University Press, Cambridge.

Morton, S. R., and Andrew, M. (1987) 'Ecological Impact and Management of Fire in Northern Australia'. *Search*, 18 (2): 77–82.

Munich RE. (1990) *Windstorm*. Munchener Ruckversicherungs-Gesselschaft, Munich.

Murty, T. S. (1977) *Seismic sea waves: Tsunamis*. Bull. 98, Fisheries Research Board, Ottawa.

Murty, T. S., and El-Sabh, M. I. (1992) 'Mitigating the effects of storm surges generated by tropical cyclones: a proposal'. *Natural Hazards*, 6: 251–73.

Murty, T. S., and Neralla, V. R. (1992) 'On the recurvature of tropical cyclones and the storm surge problem in Bangladesh'. *Natural Hazards*, 6: 275–9.

Nakamura, Y., and Tucker, B. E. (1988) 'Earthquake warning system for Japan Railways Bullet Train'. *Earthquakes and Volcanoes*, 20: 140–55.

NDO (1990) *Australian Counter Disaster Arrangements*. Natural Disasters Organisation, Canberra.

Neumann, J. (1975) 'The Mongol invasions of Japan'. *Bull. Amer. Met. Soc.*, 56: 1167–71.

Neumann, J. (1977) 'The year leading to the Revolution of 1789 in France'. *Bull. Amer. Met. Soc.*, 58: 163–8.

Neumann, J. (1979) 'The cold winter of 1657–58, the Swedish Army crosses Denmark's frozen sea areas'. *Bull. Amer. Met. Soc.*, 60: 1432–7.

Nickling, W. G., and Brazel, A. J. (1984) 'Temporal and spatial characteristics of Arizona dust storms (1965–1980).' *J. Climat.*, 4: 645–60.

Odei, H. D., Speiksma, F. T. M., and Bruynzeel, P. L. B. (1986) 'Birch pollen asthma in the Netherlands'. *Allergy*, 41: 435–41.

Oeschli, F. W., and Buechley, R. W. (1970) 'Excess mortality associated with three Los Angeles September hot spells'. *Environmental Research*, 3: 277–84.

Oliver, J. (1986) 'Natural Hazards'. In D. Jeans (ed.), *Australia, a Geography*, vol. 1 (pp. 283–314). Sydney University Press, Sydney.

Orgill, M. M., and Sehmel, G. A. (1976) 'Frequency and diurnal variation of dust storms in the contiguous USA'. *Atmos. Environ.*, 10: 813–25.

Othman-Chande, M. (1987) 'The Cameroon volcanic gas disaster: an analysis of a makeshift response'. *Disasters*, 11(2): 86–101.

Otway, H. J., and Eardmann, R. C. (1970) 'Reactor siting and design from a risk standpoint'. *Nuclear Engineering and Design*, 13: 365–76.

Palm, R. I. (1990) *Natural Hazards: An Integrative Framework for Research and Planning*. The Johns Hopkins University Press, Baltimore and London.

Pearman, G. I. (ed.) (1988) *Greenhouse — Planning for Climate Change*. CSIRO, Melbourne.

Penning-Rowsell, E. C., Chatterton, J. B., Day, M. J., Ford, D. T., Greenaway, M. A., Smith, D. I., Wood, T. R., and Witts, R. C. (1987) 'Comparative aspects of computerized floodplain data management'. *J. of Water Resources Planning and Management*, 113(6): 725–43.

Peruzza, L., and Slejko, D. (1993) 'Some aspects of seismic hazard assessment when comparing different approaches'. *Natural Hazards*, 7: 133–53.

Peterson, D. W. (1990) 'Overview of the effects and influence of Mount St Helens in the 1980s'. *Geoscience Canada*, 17: 163–6.

Pollert, S. M. (1988) 'Epidemiology of emergency room asthma in Northern California: association with IgE antibody to ryegrass pollen'. *J. Allergy Clin. Immunol.*, 82: 224–30.

Pye, K. (1987) *Aeolian dust and dust deposits.* Academic Press, London, 334pp.

Pyne, S. J. (1991) *Burning Bush: a Fire History of Australia.* Henry Holt, New York.

Quarantelli, E. L. (1993) 'The environmental disasters of the future will be more and worse but the prospect is not hopeless'. *Disaster Prevention & Management,* 2(1): 11–25.

Ramanathan, V. (1988) 'The greenhouse theory of climate change: a test by an inadverdant global experiment'. *Science,* 240: 293–9.

Raphael, B. (1979) 'The preventive psychiatry of natural hazard'. In R. L. Heathcote and B. G. Thom (eds), *Natural Hazards in Australia* (pp. 330–9). Australian Academy of Science, Canberra.

Reddaway, W. F. (1952) *A history of Europe from 1610 to 1713.* Methuen, London, 485pp.

Reid, M. J., Moss, R. B., Hsu, Y., Kwasnicki, J. M., Commerford, T. M., and Nelson, B. L. (1986) 'Seasonal asthma in northern California: allergic causes and efficacy of immunotherapy'. *J. Allergy Clin. Immunol.,* 78: 590–600.

Reiter, R. (1992) *Phenomena in Atmospheric and Environmental Electricity.* Elsevier, Amsterdam, 541pp.

Riley, S., Luscombe, G., and Williams, A. (eds) (1985) *Proceedings, Urban Flooding Conference.* Geographical Society of New South Wales, Sydney.

Rodier, J. A., and Roche, M. (1984) *World Catalogue of Maximum Observed Floods.* International Association of Hydrological Sciences, Wallingford. 354pp.

Rothermel, R. C. (1972) *A mathematical model for predicting fire spread in wildland fuels.* Research Paper INT-115, USDA Forest Service. 40pp.

Rothermel, R. C. (1983) *How to predict the spread and intensity of forest and range fires.* General Technical Report INT-143, USDA Forest Service, 161pp.

Ruckelshaus, W. D. (1983) 'Science, risk and public policy'. *Science,* 221: 1026–8.

Schlatter, T. W. (1981) 'Weather queries', *Weatherwise,* 34: 266–7.

Scruton, C. (1981) *An introduction to wind effects on structures.* Oxford University Press, Oxford.

Sedjo, R. A. (1989) 'Forests to offset the Greenhouse Effect'. *J. of Forestry,* July: 12–15.

Shears, P. (1980), 'Drought relief and agricultural rehabilitation'. *Disasters,* 4: 469–73.

Sheffield, R. M., and Thompson, M. T. (1992) *Hurricane Hugo — effects on South Carolina's forest resource.* Research Paper SE-284, USDA Forest Service, Southeastern Forest Experiment Station.

Sigurdsson, H. (1987) 'Origin of the lethal gas burst from Lake Monoun, Cameroon'. *J. Volcanol. Geotherm. Res.,* 31: 1–16.

Sigurdsson, H., Carey, S., Cornell, W., and Pescatore, T. (1985) 'The eruption of Vesuvius in 79 AD'. *Nat. Geogr. Res.,* 1: 332–87.

Simard, A. J., and Main, W. A. (1987) 'Global climatic change: the potential for changes in wildlife fire activity in the southeast'. In *Proc. Symposium on*

Climatic Change in the Southern United States (pp. 280–308). Scientific Publishing Program, University of Oklahoma.

Simkin, T., Siebert, L., and McLelland, L. (1981) *Volcanoes of the world: A regional directory, gazetteer, and chronology of volcanism during the last 10,000 years.* Hutchinson Ross, Stroudsburg, Pa.

Singer, S. (1991) 'Great balls of fire', *Nature*, 350: 108–09.

Singh, N. B., Taylor, P., Bellomo, R., Holmes, P. Puy, R., and Knox, R. B. (1992) 'Mechanisms of grass pollen induced asthma'. *Lancet*, 69: 569–72.

Skaf, R., Popov, G. B., and Roffey, J. (1990) 'The Desert Locust: an international challenge'. *Phil. Trans. R. Soc. Lond.*, B328: 525–38.

Smith D. I. (1990) 'The worthwhileness of dam failure mitigation: an Australian example'. *Applied Geography*, 10: 5–19.

Smith D. I. (1993) 'Drought policy and sustainability: Lessons from South Africa'. *Search*, 24: 292–5.

Smith, D. I., Handmer, J. W., Greenaway, M. A., and Lustig, T. L. (1990) *Losses and lessons from the Sydney floods of August 1986.* Mimeo, 2 volumes, 121 & 98pp. CRES, Australian National University, Canberra.

Smith, J. B., and Tirpak, D. (eds) (1988) *The potential effects of global climate change on the United States.* US Environmental Protection Agency, Washington, DC.

Snyder, C. H. (1991) 'That's the way the building crumbles'. *Weatherwise*, 44(3): 28–32.

Stark, K. P., and Walker, G. R. (1979) 'Engineering for natural hazards with particular reference to tropical cyclones'. In R. L. Heathcote and B. G. Thom (eds), *Natural Hazards in Australia* (pp189–203). Australian Academy of Science, Canberra.

Steadman, R. G. (1971) 'Indices of windchill of clothed people'. *J. Appl. Meteorol.*, 10: 674–83.

Steadman, R. G. (1979) 'The assessment of sultriness. Part I: A temperature-humidity index based on human physiology and clothing science, and Part II: Effects of wind, extra radiation and barometric pressure on apparent temperature'. *J. Appl. Meteorol.*, 19: 861–84.

Stretton, A. B. (1976) *The Furious Days.* William Collins, Sydney.

Stretton A. B. (1979) 'The role of the Natural Disasters Organisation'. In R. L. Heathcote and B. G. Thom (eds), *Natural Hazards in Australia* (pp. 359–73). Australian Academy of Science, Canberra.

Swanson, S. E., and Kierle, J. (1988) 'The 1986 eruption of Mount St Augustine'. *J. of Geophysical Research*, 93(B5): 4500–20.

Symmons, P. (1992) 'Strategies to combat the desert locust'. *Crop Protection*, 11: 206–12.

Tannehill, I. R. (1947) *Drought, Its Causes and Effects.* Princeton University Press, Princeton, NJ.

Taylor, G. A. M. (1958) *The 1951 eruption of Mt Lamington, Papua.* Bull. 38, Aust. Bureau of Mineral Resources.

Thieler, E. R., Bush, D. M., and Pilkey, O. H. (1989) 'Shoreline response to Hurricane Gilbert: lessons for coastal management'. In O. T. Magoon et al.

(eds), *Coastal Zone 89* (pp. 765–75). American Society of Civil Engineers, New York.

Tiedmann, H. (1990) *Newcastle: the writing on the wall.* Swiss Reinsurance Company, Zurich.

Tilling, R. I. (1989) 'Volcanic hazards and their mitigation: progress and problems'. *Reviews of Geophysics,* 27: 237–69.

Time-Life (1989) *The Age of Calamity: Time Frame AD 1300–1400.* Time-Life Books, Alexandria, VA.

Titus J. G. (1988) 'Sea level rise'. In Smith, J. B., and Tirpak, D. (eds) *The potential effects of global climate change on the United States* (pp. 9-1 to 9-47). US Environmental Protection Agency, Washington, DC.

Todd, B., and Kourtz, P. H. (1991) *Predicting the daily occurrence of people-caused forest fires.* Information Report PI-X-103, Forestry Canada, Petawawa National Forestry Institute.

Touliatos, P. (1993) 'The traditional aseismic techniques and the everlasting principles they reveal'. *STOP Disasters,* 12: 4–5, plus supplement: 1–8.

Tuan, Y. F. (1974) *Topophilia: a study of environmental perceptions, attitudes, and values.* Prentice-Hall, Englewood Cliffs, NJ.

Tuan, Y. F. (1979) *Landscapes of fear.* University of Minnesota Press, Minneapolis.

UNDRO (1988) 'Resolution on the International Decade on Natural Disaster Assistance', *UNDRO News,* January/February.

UNEP (1991) *Environmental Data Report, 1989/90.* Basil Blackwell Ltd, Oxford, for United Nations Environment Program.

van Wagner, C. E. (1987) *Development and structure of the Canadian forest fire weather index system.* Forest Technical Report 45, Canadian Forestry Service, 37pp.

Vasconcelos, M. J., and Guertin, D. P. (1992) 'FIREMAP — Simulation of Fire Growth with a Geographic Information System'. *Int. J. Wildland Fire,* 2(2): 87–96.

Walker, G. R. (1990) *Report on Newcastle Earthquake, Australia.* DBCE DOC 90/11(s), CSIRO, Division of Building, Construction and Engineering.

Wallace, G. (1993) 'A numerical fire simulation model'. *Int. J. Wildland Fire,* 3: 111–16.

Walsh, J., and Warren, K. (1979) 'Selective primary health care: an interim strategy for disease control in developing countries'. *New England J. of Medicine,* 301: 967–74.

Watson, A. (1992) *An exceptional Adelaide hail-storm and the associated severe convective outbreak: 22 and 23 January 1991.* Technical Report 65, Bureau of Meteorology, Melbourne.

Watt, M. (1983) 'On the poverty of theory: natural hazards research in context'. In K. Hewitt (ed.), *Interpretations of Calamity.* Allen and Unwin, Boston.

Webb, G., and Manolis, C. (1989) *Crocodiles of Australia.* Reed, Sydney.

Wheaton, E. E. (1992) 'Prairie dust storms — a neglected hazard'. *Natural Hazards,* 5(1): 53–63.

Wheaton, E. E., and Chakravarti, A. K. (1987) 'Some temporal, spatial and climatological aspects of dust storms in Saskatchewan'. *Climatol. Bull.*, 21(2): 5–16.

Wheaton, E. E., and Chakravarti, A. K. (1990) 'Dust storms in the Canadian prairies'. *Internat. J. Climatol.* 10: 829–37.

White, G. F. (ed.) (1974) *Natural Hazards: local, national, global.* Oxford University Press, New York.

White, G. F., and Haas, J. E. (1975) *Assessment of research on natural hazards.* MIT Press, Cambridge, Mass.

Whittingham, H.E. (1963) 'The Bathurst Bay Hurricane'. *Austr. Met. Mag.*, 42: 37.

Wijkman, A., and Timberlake, L. (1984) 'Natural Disasters: Acts of God or Acts of Man?' *Earthscan*, International Institute for Environment and Development, London and Washington, DC.

Wilhite, D. A. (1986) 'Drought policy in the US and Australia: a comparative analysis'. *Water Resources Bull.*, 22: 425–38.

Wilhite, D. A., and Easterling, W. E. (eds) (1987) *Planning for Drought: Toward a Reduction of Societal Vulnerability.* Westview Press, Boulder, Colorado, 597pp.

Wilhite, D. A., and Easterling, W. E. (1989) 'Coping with drought: toward a plan of action'. *Eos*, 70: 97, 106–8.

Wilkie, W. R. (1984) 'The Australian tropical cyclone warning organisation'. *Disasters*, 8(2): 143–8.

Williams, R. S., and Moore, J. G. (1973) 'Iceland chills a lava flow'. *Geotimes*, 18: 14–17.

Williams, S. J., Dodd. K., and Gohn, K. (1991) *Coasts in Crisis.* Circular 1075, Geological Survey, US Dept. of Interior, Reston, Va.

Wilson, G. (1992) *Pest animals in Australia: a survey of introduced wild animals.* Bureau of Rural Resources/Kangaroo Press, Kenthurst, Sydney.

Wood, F. J. (1976) *The strategic role of Perigean Spring Tides in nautical history and North American coastal flooding, 1635–1976.* National Oceanic and Atmospheric Administration, Washington, DC.

Yevjevich, V., Da Cunha, L., and Vlachos, E. (eds) (1983) *Coping with Droughts.* Water Resources Publications, Colorado.

Index